C. B. RADIOS
A Practical Guide

C. B. RADIOS
A Practical Guide

By
MADELINE DOLOWICH

GALAHAD BOOKS   NEW YORK CITY

Library of Congress Catalog Card Number: 76-16712
ISBN: 0-88365-371-0

Published by arrangement with Drake Publishers Inc.
801 Second Avenue, New York, N.Y. 10017

Manufactured in the United States of America

Simultaneously published by Drake Publishers Inc.
under the title of THE COMPLETE BOOK OF C.B. RADIOS

CONTENTS

DEDICATION

An author usually accrues lists of people to whom dedications are due. This writer is no different. With much love and devotion, this work is dedicated to Buddy, Jodi, Robin, my parents, my brother- and sister-in-law, their son, and anybody else I may have inadvertently forgotten.

MADELINE DOLOWICH

C. B. RADIOS
A Practical Guide

CHAPTER 1

INTRODUCTION

You have taken the bother to pick up this book and leaf through its contents. Good! It means that you are, first of all, interested in the above-mentioned subject, and that, secondly, you are probably an individual who is constantly wondering at what makes things tick. This also means that you have just taken your first step toward learning about a media which is growing in popularity by leaps and bounds. Citizens Band radio!

This book is designed for the layman, to inform, not to confuse, and will be written in plain English, and will not be cluttered and bogged down with uninteresting, heavy text words.

I have always been interested in mechanically run operations—how I became interested in C.B. radios some years back, was simply due to a quirk of fate.

It all started in my friend's car one day. How many other times I had been in it, I don't know, but although I meant to ask her time and time again about that impressive looking gadget resting

1-1 A typical multi-channel C.B. transceiver, ready to be installed.
COURTESY E.F. JOHNSON CO.

just below her dashboard, we always launched into conversation the way friends often do, and my attention veered once again from the object.

But this time, as we were about to pull out of her driveway, the car would not start. Try as she may, nothing happened when she turned the key in the ignition.

Before I knew it, she picked up something which I later learned was called a *microphone*. Suddenly, she began to talk into it and another voice answered her back.

"What's that thing?" I asked, a flicker of interest beginning to tug at me.

"A Citizens Band radio," she answered.

"A what?" I responded.

But before I could question her further, she was explaining to the voice at the other end about her car trouble.

That night, I told my husband what had happened during the day and was surprised to find out he had one of these units in his office.

Little by little I discovered that over four million people use Citizen Band radios as part of their daily lives. I was really surprised when I found out that a cousin of mine living half-way across the United States was a "nut" on the subject.

When I saw my friend again, I told her how interested I had become in the radio. With that, she took me into her kitchen and showed me the unit she kept on top of her counter. It was black. Small. Hardly taking up any space in a tiny corner where she had it plugged into a wall socket.

"Show me how it works," I said, suddenly excited by it.

She picked up the trim microphone and horned in on a conversation between two men. My friend identified herself and broke into their conversation. There seemed to be an air of friendliness about the whole thing.

C.B. radio users are very courteous to each other because they know there is a five minute limit on how long you can talk over the unit. You are supposed to say what you have to say and then get off and give somebody else a chance to talk.

Using foul language over the unit is against the law and, because the FCC does listen in, you can get caught. The penalty can be a rather stiff one, too.

The Federal Communications Commission organized Citizens Band radio when amateur radio operators abandoned what they called the eleven-meter band. Now, to get it straight, a C.B. radio is not like a ham radio or a commercial radio with which you would tune into your favorite rock station. No. What it actually is, is a two-way radio, very much like a party-line telephone system.

The difference is that the C.B. radio works on radio waves instead of telephone lines. In fact, you can usually recognize a C.B.-equipped automobile by the antenna sitting on top of the back of the vehicle.

1-2 Before installing the antenna atop the house, first it must be assembled. The best place to do such assembly work is not on the roof, but on the ground.

When the FCC gave this band to the citizens, there were certain stipulations: the station had to be licensed by the FCC; it had to be of an approved type so that there would be no interference with other services; and it had to be limited to five watts of power.

Anyone except an alien of the United States can obtain a license. Usually, there is a long delay in getting it because there are so many applicants. You do not need a license for every unit you own. No matter how many you own, all can work on the one license.

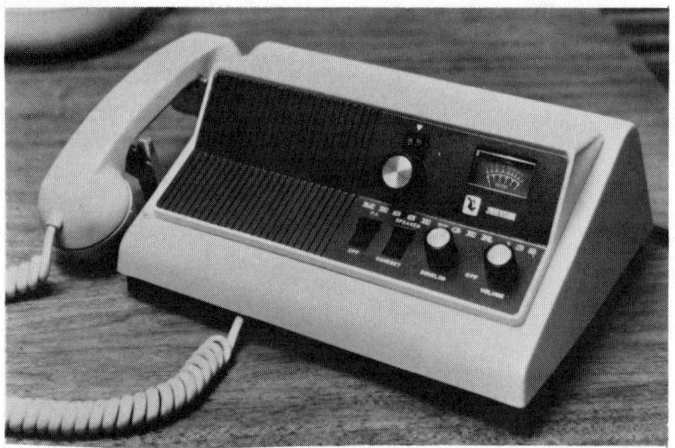

1-3 Equally at home in the car, on a kitchen table or on a desk, as seen here, the C.B. transceiver is as handy as a phone!
COURTESY E. F. JOHNSON

Power Rating and Kinds of Units Available. There are small portable units that resemble toys. You may have seen kids running around with them. These plastic models sell at very low prices and need no license because it runs on under one-hundred milliwatts. They have a range of from twenty-five to fifty feet.

You do need a license, though, if your unit operates between one-hundred milliwatts and five watts. Remember, you cannot go higher than five watts.

What exactly is a *watt?* A watt is the amount of power generated when one *volt* of electric force pushes one *ampere* of electric current through one *ohm* of resistance.

The volt is a unit of force of electricity and the amp is the unit of electrical capacity. If you recall, you usually have one-hundred and ten volts in your house.

With the five-watt power limitation, C.B. users thought they would never be able to get out of their backyard. But they did. They figured that if they could not increase the strength of the transmitter by adding power to it, at least they could put in a better receiver and be able to hear further distances away.

So less expensive units have what is called a *regenerative receiver.* What it does is receive signals but it will get interference from other channels. Neither is it *tunable selective* and it has a tuning dial instead of a switch to switch from channel to channel with.

More expensive models are more sensitive and more selective. *Sensitivity* in a receiver means how weak a signal it could pick up and amplify. *Selectivity* refers to how close a station can be and still not be heard on the receiver.

1-4 Citizens Band Radio being used in its early stages.

Another feature of the C.B. radio is the *squelch control*. This suppresses static noise. Some of the more expensive units use a *caller* device instead. For instance, if you have a transmitter in the house and one in the car and your husband presses the button on the microphone, you will still hear a squeal in your receiver even if it is turned off. That means you should tune in and listen as he is about to talk.

There are twenty-three channels. Channel nine is used for emergency calls. Not all *transceivers* come equipped with every channel (the transceiver is a transmitter and receiver mounted on a single chassis). You could buy a unit that just gets channel nine and use it only as an emergency device. Having just one channel makes it less expensive than other radios.

There are other units that come with four channels. They are lettered A, B, C, and D. Different crystals are plugged in to determine what channels you wish to hear.

Each channel has two sidebands: an *upper* and a *lower*. Together they make up AM radio, which is what Citizens Band operates on. If you have what is called a *sideband rig*, you can throw a switch to use either of the sidebands and communicate without interfering in any of the other stations on that channel. Sometimes this is also called a *single sideband* because of the fact that it operates on the upper or lower sideband. Since there is limited space, this is a good way to double the space—Citizens Band radio is growing so fast in popularity, creating the possibility of overcrowded bands in the future.

A radio *antenna* is a conduct or device for receiving or for broadcasting radio waves. The British, and a few Americans call them aerials.

The kind of antenna in use today, mostly on cars, is the straight type called a *whip*. The whip is about eight feet long. There are also smaller ones, called *loaded antennas*, that have little coils in the middle. They are more attractive in appearance

1-5 Typical Citizens band base station installation on a residential home.
COURTESY OF NEW-TRONICS CORP.

but not quite as efficient. Portable C.B. units have collapsing antennas. Since they are not on all the time it is not necessary to have an eight foot antenna sticking out of a six inch transceiver. These collapsing antennas slip down into the case and stay out of sight. There are better things you can do with a fixed-station antenna (such as in your home) because it will not have to be mounted on a moving vehicle.

Some Important Uses for C.B. Radio. This matter is taken up in Chapter 2, but the following examples serve to point up some of the more important uses for C.B. radio. These are practically limitless. For instance, two families could be going on a trip together, each driving their own C.B. radio-equipped car. Car A is supposed to lead the way. Car B will follow because he doesn't know the way and is depending on Car A's instructions. While they are driving, Car A loses sight of the other car. What does he do? He communicates with the other driver by way of C.B. radio. He can ask where he is supposed to make that all important turn.

During the winter many of the ski resorts in the northeast make use of C.B. radio on certain channels—even for advertising. So if you happen to be going to a ski resort and you see that the roads are not very good, and by now you are really cold and tired, you can pick up your microphone to call the desk clerk. He will reserve a room for you and have a hot meal ready for you upon your arrival. Not only that, he could tell you which roads are in the best condition, and give you instructions as to how to reach the resort.

Parents can give their small children little portable units to keep in their pockets. In this way, the parents can contact them at any time.

Do you own a dog? C.B. radio can be of assistance. By putting a small unit on your dog's collar, you can let him run free for his exercise. He'll come bounding home when he hears your familiar

1-6 Delux dual rain gutter C.B. antenna installation.

whistle over the unit. Just imagine a cold day that looks as if it might rain or snow any minute, and you are in a nice warm house, all snug and cozy and lazy, and your dog begins to whimper because he has to go out. Now there doesn't have to be any more family arguments over who will take Rover out. Just let him go out by himself and whistle into your transmitter to bring him back.

A friend of mine who likes to ski went to take some lessons. She told me that the instructor used some kind of device for her lessons. Know what it was? Right! You guessed it. A Citizens Band radio. The instructor put a small unit in the pocket of her jacket while he had a unit for himself at the bottom of the slope. While my friend was swooshing down that slope, he gave her instructions on how to correct her errors while she was in the process of her stint.

Truckers use it to help each other keep awake on long hauls, and to warn one another of road hazards ahead.

A husband can use C.B. radio to contact his wife. For example, if he should get caught in traffic jam, he can signal her to stop cooking because he got tied up in heavy traffic. So she turns off her oven and saves her food from burning and also saves herself the worrying she ordinarily would have gone through if she did not have direct communications with her husband.

So you see, Citizens Band radio has many uses. People now feel that their unit is just as important to them as their telephone; they like the fact that it is so durable and compact, and that it could be installed practically anywhere.

Doctors use it, printers, students, lawyers, housewives—almost anyone in business. Users can talk to or from home, to or from cars, business, or even to private aircraft and boats.

Citizens Band radio is also used by magicians in their act. They use it as a signaling device. Something called a *relay* is connected to a receiver. Then it is connected to a flash pot, a small metal box with flash powder in it. The magician will either have the transmitter in his pocket or somewhere off stage. He simply presses the button and there goes the flash.

The television and movie industries use C.B., too. As you know, a microphone can pick up sound from just so far away. But think about this.

How many times have you seen a movie where in one scene the actors were at least one-hundred feet away and you were still able to listen to their conversation? They had small microphones built into their clothing and connected to a C.B. transmitter. The receiver remains back with the audio equipment and the camera operator. You've seen this time and time again on television where there were no microphones. You were always able to hear the sound going on in the show.

Transmitting and Receiving on C.B. Radio. In Citizens Band radio, when you push the microphone to talk, and release the mike button, it automatically switches to receive.

However, there is yet another type of operation called VOX. When you talk the sound of your voice automatically turns the transmitter to transmit. When you stop talking the silence puts it back on receive.

Another feature of this type is called *duplex operation.* You can have your transmitter and receiver on at the same time. The other party does likewise. You talk on one channel and he listens to the channel on which you are talking. Then he talks on the channel on which you are listening. What happens here is that you can both talk and listen at the same time. It's almost like talking on the telephone.

Usually, people have to adjust themselves to using C.B. radio, because they are not used to radios. They have to try and remember that when they touch that microphone button, the transmitter is on. When the transmitter is on the receiver is then blocked. You simply cannot hear. So when you are finished talking and you want to be answered, you have to release the button so the other party will be able to talk. When the party is through talking he or she, in turn, will also have to release the button so you can talk again and you

1-7 High efficiency 30″ trunk lip mount C. B. antenna.
COURTESY NEW-TRONICS CORP.

must keep in mind to press your button down. Otherwise, if you forget to do so, your party will not be able to hear you.

Applying for a C.B. Radio License. The application for a license used to be pretty complex and many applications were rejected by the FCC because they were not filled out properly. The pity of it was that you had to wait for months—and you still do—to get your application processed and it is really very disheartening to have it come back to you without a license because something was not filled out correctly. There is no other choice but to

1-8 Dual trunk lip mount C.B. installation antenna.
COURTESY NEW-TRONICS CORP.

fill out the new application and start waiting all over again. Later, we will discuss how to fill out an application correctly. Some changes in the forms have made it a more simple task. But isn't it a good thing that you only need the one license to cover the amount of C.B. radios you own?

You know, with ham radios, it's all very different. Each person must own a separate license and only the licensed operator is allowed to operate a station. Also, you have to go through difficult tests and even learn Morse Code.

And so, we are on our way to learning about . . . what do you call that thing again? . . . oh, yes . . . Citizens Band radio. . . .

CHAPTER 2

WHY CITIZENS BAND RADIO?

There are three good reasons for having Citizens Band radio:

SAFETY
CONTROL
CONVENIENCE

Picture this: A man is sound asleep in his own home. Suddenly, a noise wakes him up. He realizes at that precise moment that his house is being burglarized. Luckily, his bedroom door is shut so the burglar, who is so busy with his looting, doesn't hear his awakened victim. And now, very quickly, the man reaches over for the telephone standing on his night table beside his bed. He is aware that he could be caused bodily harm if the intruder should decide to come upstairs. He begins to dial. But he finds out that the phone is dead. The robber had severed the wires. However, all is not lost. He tunes in to channel nine, the emergency channel on his Citizens Band radio and sends out a call for help.

Only precious minutes later, the police arrive on the scene to arrest a very surprised and puzzled thief. After the looter had been removed from the premises, the police informed the intended victim that after they received his plea for help, other C.B. users who had been listening in, heard his cry and bombarded police headquarters for five minutes after he had placed his call.

Now, rest assured, that this man, never for one moment, thought when he bought himself a Citizens Band radio, that it would ever be used one day in the face of an emergency, and would perhaps, save his life.

In the early 1970s, there was a gas shortage and a truckers' strike, both occurring at the same time. A new speed limit was set at fifty-five miles an hour. The oil embargo hit the truckers in a crossfire of higher fuel prices and lower earnings. They struck back, using C.B. to tie up traffic in major cities. But then every long-haul trucker saw another way to use the radio. They warned each other of speed traps and road hazards.

I, myself, was saved from a ticket for speeding, even though I didn't have a C.B. in my car at that time. I was on a vacation half way across the United States, driving along a main highway when a big truck in front of me was bogging my progress down. I tried to pass it but the truck just drifted into the passing lane and blocked me. Startled, I again attempted to pass the creeping giant but it continued to block my way.

By this time I really began to get angry, but my anger didn't do me any good, because pretty soon I saw that despite my attempts to pass, I had no choice but to follow behind at the exact rate of fifty-five miles per hour. Oh, how I wished the road would widen so I could pass him by, or that perhaps somewhere along the line he would turn off. None of these things happened. He stayed in front of me all the way.

But about a mile later we passed a well-concealed radar trap. I was instantly reminded that the truck had been equipped with a C.B. unit and I was not. I could tell he was C.B.-equipped by the antenna he had. Through it, his fellow truckers had warned him about the speeding trap. Anyway, just as we both passed the trap, he waved for me to go ahead of him and pass.

There are many applications for C.B. radio that go beyond the requirements of communication. It is practically impossible to list all of them.

Interestingly enough, did you know that people who fly model airplanes use C.B. radio? The model planes are radio-controlled. They send out pulses and every time the receiver up in the airplane receives a pulse, it alters the control slightly, just enough to make the plane maneuver.

Citizens Band radio can be used to operate garage doors, too.

COURTESY SURVEYOR MANUFACTURING CORP.

A magician can use a C.B. unit wrapped in his turban while he is performing on the stage. His assistant, usually some gorgeous young thing, has the transmitter somewhere on her person. She will go out into the audience and from there pick a member of it to whisper a number in her ear. While the person whispers it, the transmitter relays it to the magician who hears the number on the receiver hidden in his turban. And then, poof! He repeats the magic number and takes his bows.

Boats have what is called a marine radio telephone. With it, people talk to other boats and the Coast Guard along the shore. Otherwise, they would have to communicate with signal flares. Can you picture that? Every day would look like the Fourth of July.

It would cost about three-hundred to five-hundred dollars to install a marine radio in your boat. But if you want to install a Citizens Band radio it would cost under one-hundred dollars. Why? Because a popular, mass-produced product that is selling well can be sold in greater quantities at a reduced price. Savings are passed along to the customer. Besides, knowing that you are equipped with a C.B. radio, and that you are not limited to talking with other boats or to those along the shore, but can communicate with thousands of other people as well, certainly gives a great sense of security.

Many pilots use C.B. in their private airplanes just for added safety. Instead of being limited only to talking to the tower or to other planes with their own radio equipment, they can be just as secure as are the boat owners.

C.B. can be used in other ways, too. It can be used in cars, in homes, or as a portable in a hand-carry unit. There isn't anyone you can't talk to. With a portable, you could talk to your child on the street—e.g., to call the children home to dinner.

According to the rules governing the use of C.B. radio, it is self-monitoring. This means that you are not permitted to complain to the FCC that you were talking and someone interrupted. People are allowed to do just that. So there is a way that C.B. users handle this. With common courtesy.

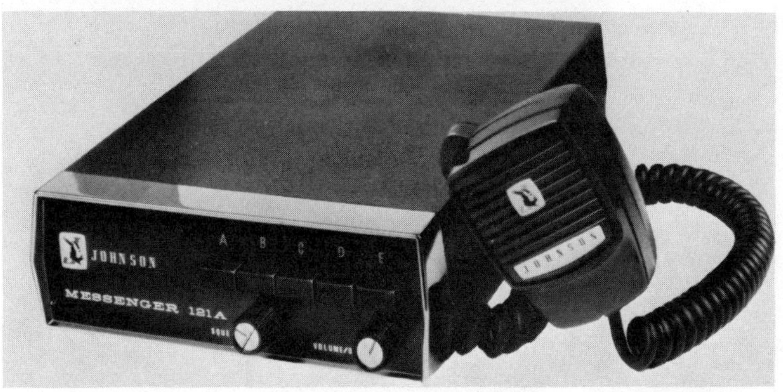

COURTESY EQ. F. JOHNSON COMPANY

Someone I know who was once driving along a road in upper New York state with his C.B.-equipped car got the signal to beware of cows blocking the road. He was driving at seventy miles per hour when he got the call to slow down. It was lucky he had the C.B. rig because it may very well have saved his life. About a mile and a half later he came to a curve in the road and there, standing directly on the road was a big herd of cows. Because of the warning, he had slowed

down enough to safely come to a full stop. Had he
not received the alert, he probably would have
crashed into the cows killing some of them along
with himself.

When you begin to think of owning a Citizens
Band radio, you might think you want to buy it for

COURTESY SURVEYOR MANUFACTURING CORP.

one specific purpose. That's fine. But later on,
you'll keep finding more and more ways to make it
more handy for you. An emergency is usually the
last thing on anyone's mind when they buy the rig.
However, it could very well save your life one day.

In many locations, C.B. radio has become the
new backyard fence for ladies. Yes. They can
turn on their sets, turn the volume up and tune into
conversations while going about their business.
When it's their turn to talk, they'll take a break and
talk for a minute or two. C.B. radio has become a
modern and interesting way to find out where the
best sales are in town for buying meat, dresses, or
whatever.

Citizens Band radio plays a very important part

COURTESY TENNELEC, INC.

in the average American home. There are quite a number of people who still do not drive cars. Because of the isolation such people depend very heavily on C.B. radio as a means of communication with one another. In this way, with the convenience of owning a unit, they can make do without a telephone. Thus, Citizens Band radio is both a convenient and inexpensive way of communication.

Monitor-Scanner. This is a device which automatically checks each channel. If you want to talk to somebody and you don't know who is on the air, you just push the scanner button. It stops when it hears somebody talking. This is a quick way of checking when you are driving.

The JMD Electronics SMS series monitors are ten-channel, crystal-controlled, solid state monitor scanning receivers which let you monitor up to ten police, fire, or public safety channels automatically. This monitor scanning receiver checks each of its channels, one at a time, and stops at the first channel on which it picks up a signal on. It will stay on that channel, until it stops receiving the signal and will then recheck the channels again for another signal.

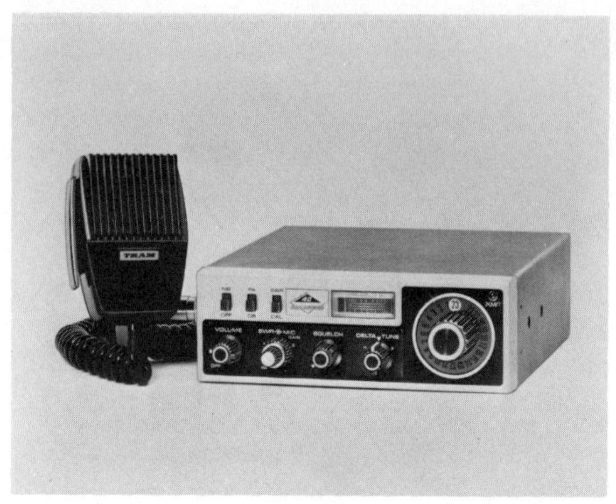

COURTESY TRAM CORPORATION

There are ten control switches mounted on the front panel to enable the user to shut off individual channels so they will not be checked. The switches can be used in scan position for automatic checking or in manual position. Up to ten crystals can be easily installed. The unit is available in single or dual band. In the dual band receiver any combination of high and/or low crystals can be installed in any order. The monitor is "double conversion" to suppress unwanted channels and images. It has a solid state receiver, requires no warm up time and because of the receiver's low current drain, it allows listening for extended periods of time in an automobile with the engine off. A whip antenna comes with it for general use. It also has a built-in four-inch speaker with a jack for an external speaker. Attractive with its wood grained vinyl cabinet with black and silver front, it can be used in the home, using the A.C. power cord or may be used in an automobile by using model MTB bracket and D.C. cord. The receiver checks each channel that is turned on and stops at the channel on which it detects a signal.

Possible Emergency Uses for C.B. Radio. Citizens Band radio serves an important purpose in suburban areas. For example, if you live in an area where you don't have one small town after another but have desolate highways instead, and you drive mile after mile and the area is unpopulated—isolated, C.B. can be of value in lessening this isolation, or in serving to communicate with others in an emergency.

Suppose that you are an attractive woman, and for some reason you become stranded. You need help. What could you do? The police advise you to tie a white handkerchief on your door handle. It is the universal appeal for help. A car might stop, sooner or later. A man could get out and come up to you. But your heart is in your mouth. Is he going to help you with your car or is he going to help himself to you? You are at his mercy and the patrol cars don't seem to come around that often.

If you had a C.B. radio in your car you could get help the safe way—and fast. Or, if you didn't want anyone to come to your assistance, you could call in on the emergency channel and relate the message that you'd like someone to contact your husband or a neighbor to come and help you. This shows that C.B. provides a margin of safety in such emergencies.

Suppose your car broke down in a dangerous neighborhood of a city? Even if you are a man you could be stooped over, working very hard at your task, but looking anxiously up at every passer-by. It might happen late at night. A cat could jump out of a garbage can and frighten the daylights out of you. And who are those weird looking guys in leather jackets? What's in their pockets? Guns? Knives? Nothing?

Again—C.B. to the rescue. A simple call over your unit could bring you comfort, safety, and assistance. You never know when an emergency will arise. Any kind. That's why you carry Blue Cross or Blue Shield. You never expect to need it, but you want to be prepared for an emergency.

The Origin of C.B. Radio. Actually, Citizens Band radio started off rather quietly. Sometimes people confuse C.B. radio with amateur radio, but the two are completely different radio services. They tend to be confused because both are used by private individuals, rather than by corporations or governments. Amateur radio operators must pass hard government examinations to get their licenses. Citizens radio operators merely fill out a single form.

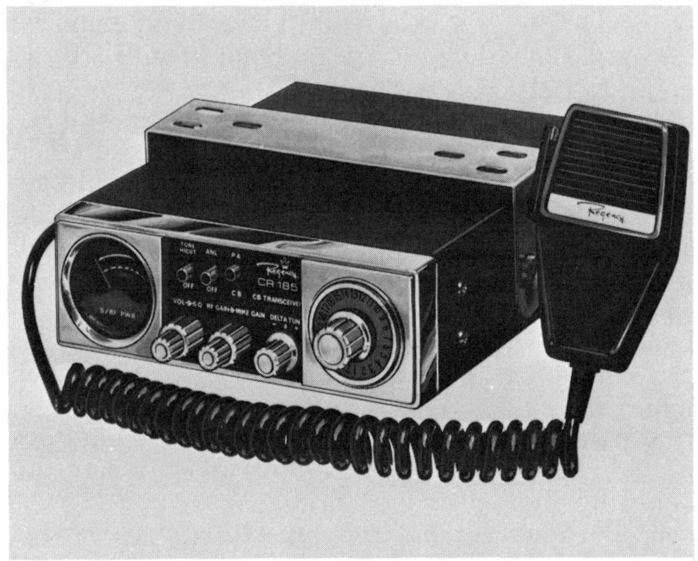

COURTESY REGENCY ELECTRONICS, INC.

Amateurs have been assigned ten communications and twelve experimental radio bands and can use one-thousand watts of transmitter power. Citizens radio operators can use only one communications band, and can use only five watts of transmitter power.

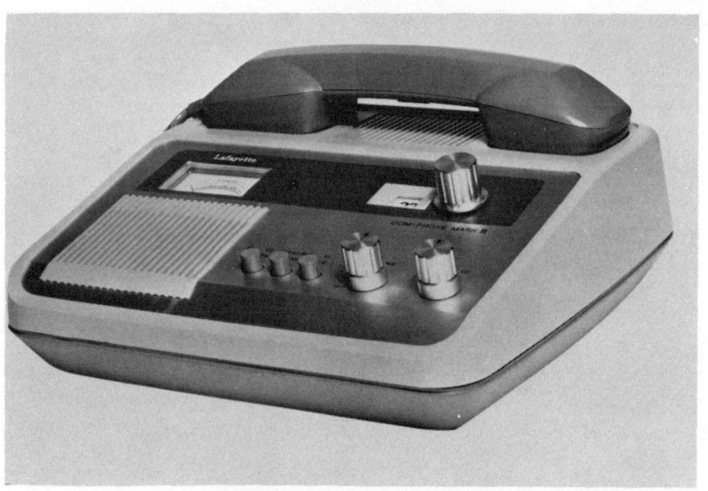

COURTESY LAFAYETTE RADIO ELECTRONICS CORPORATION

The legal purposes of amateur radio are providing emergency communications in time of disaster, advancing the engineering and technical aspects of the science of radio communication, broadening radio skills among the general population, and enhancing international goodwill. Amateur radio is specifically recognized in law as a hobby involving long range communications. The legal purpose of citizens radio is "short-distance personal or business radio communications."

In the 1950s, the Federal Communications Commission decided there should be a place in the radio spectrum where ordinary citizens could use radio for short-range personal and business communications. "Short range" was defined as less than one-hundred fifty miles and within a single metropolitan area. Then in 1958, the FCC opened the **Class D Citizens Radio Service** on an eleven-meter band.

The expression, "eleven-meter band" needs explanation. A radio wave travels 186,000 miles per second. C.B. radio is authorized to operate around

23

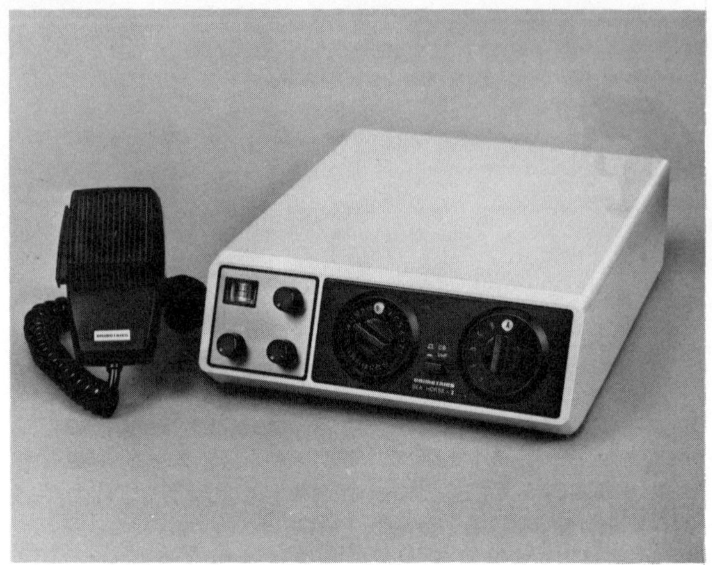

The Citizens Radio Service, as it is officially called, started off quietly. Few people had heard of it, and it was a delight to those who pioneered it in the beginning. It was simple to use, easy to get a license, and communications were interference-free.

By 1960, the teenagers and pre-teens in the cities had discovered it. After-school chatter became incessant. In the evening the adults took over while the kids did their homework. By 1961, the kids in the big cities were taking up all the talking time over all the channels. The adults responded by getting bigger transmitters. But these are now illegal.

The kids often got hold of these bigger transmitters and an unbelievable power competition ensued. Some stations actually had several thousand watts of power, while a great many were running several hundred watts. The fascination with power led to working long distances.

To avoid detection, C.B. radio users stopped using their FCC assigned call signs such as KBG-0505, and became known by self-bestowed nicknames such as Big Red, or Poppa Bear, Night Rider, etc. Today these wonderfully colorful nicknames are used in every corner of the United States.

The Federal Communications Commission had never actually stopped to handle this sort of thing. They made a few arrests and wrung their hands. Eventually, they even made a few minor alterations in the rules and regulations. But pretty soon every FCC field engineer was completely swamped in a tidal wave of Citizens Band users. The whole thing flared into a friendly, illegal bedlam and kept ballooning even more.

However, as of today, Citizens Band radio is booming. Every new C.B. radio user that comes along reaps the benefits of his or her personal radio. C.B. radio has opened up a whole new market.

More and more, you see vehicles equipped with a C.B. rig. The telltale antennas are easy to spot. And, more and more, you find that many more people have C.B. radios in their homes.

CHAPTER 3

WHO IS ELIGIBLE AND WHAT ARE THE REQUIREMENTS?

If you are an adult over the age of eighteen and you have the need for communications, then you are an eligible candidate for a license. You must be a citizen of the United States and not liable for any crimes committed, for criminals give up their rights to citizenship. The license fee used to cost as much as twenty dollars, but is now only four dollars.

The Federal Communications Commission has laid down a number of rules. For example, you cannot build your own unit, no matter how handy you might be. This is so because your transceiver must be of an approved type that will not interfere with other services such as ambulance, police, fire, etc. By having the manufacturer submit a sample unit for periodical testing, the necessary standards can be maintained. When you buy your unit, notice that it says FCC type approval for service on the back of it.

One problem that has arisen out of Citizens Band radio is its easy availability. People who didn't understand thought they could go into a store and buy a unit and start talking on it immediately. They thought they could just get on the air without a means of identification. At one time you had to give your license number when you got on the air, but now you can give your "handle" or code name.

Until recently, application for a license required filling out a many-page form. In December, 1974, the FCC came out with a new 505 form which consists of a single page. Here is how to fill out the new form:

At the top of it are instructions. It tells you to use a typewriter or to print clearly in capital letters and to stay within the boxes. Then you are to skip a box where a space would normally have appeared.

Now, on the first line of the form 505, you are to state your name and date of birth. If your license is for a business, fill in the name of it and be sure to include the address.

Note numbers eleven, twelve, thirteen, and fourteen and make certain you check the appropriate box. Notice that number fifteen certifies the following: That the applicant is not a foreign government or a representative thereof. That the applicant has (or has ordered from the Government Printing Office) a current copy of Part 95 of the Commission's rules governing the Citizens Radio Service. That the applicant will operate his transmitter in full compliance with the applicable law and current rules of the FCC and that his station will not be used for any purpose contrary to federal, state, or local law or with greater power than authorized. And finally, that the applicant waives any claim against the regulatory power of the United States relative to the use of a particular frequency or the use of the medium of radio waves because of any such previous use, whether licensed or unlicensed.

You will notice that it says willful or false statements made on the form or attachments are punishable by fine and imprisonment. Sign your name to this statement and also put the date down.

Next, sign and date the application. Then enclose your fee with the application. You are not to submit cash. Make out a check or money order payable to the Federal Communications Commission. The form goes on to tell you that no fee is required for an application filed by a governmental entity. And for additional fee details, including the amount and exemptions, see Subpart G of Part I, FCC Rules and Regulations. You are not to enclose the order form or subscription fee for FCC

Rules. Then mail your application to the Federal Communications Commission, Gettysburg, Pa., 17325.

On the bottom of form 505 is the order form for your subscription to obtain Parts 95, which are very pertinent to a future Citizens Band radio user, and also Parts 97, and 99 of the Federal Communications Commission Rules and Regulations. Make sure you enter your name, address, and zip code properly. Mail your order form to the Superintendent of Documents, Government Printing Office, Washington, D.C. 20402.

The application form is usually included in the instruction book and the guarantee and the whole package you get when you buy your unit. Filling out the form is not in itself sufficient. There is usually a waiting period—sometimes a long one—due to the boom of people waiting for their licenses, too. You must obtain a validated license before operating your rig.

Do not, under any circumstances, operate your C.B. radio without your license. This cannot be stressed too heavily. You will only be caught and have to pay a severe penalty.

Children are not eligible. Confined and handicapped people are eligible. This includes persons who are confined to wheelchairs or who are bedridden or crippled and cannot get out of the house. They might not have any way of communicating with people outside their immediate families. For such persons C.B. radio opens up a whole new way of life.

You know, sometimes people seem kind of wary when they have an amateur radio operator in their neighborhood. This could be because they think that if you have an antenna on the top of your roof, it will interfere with their television reception. Not so. It is not so because the frequency of the television is so far from the frequency of Citizens

Band radio that there simply cannot be any problem.

Extracts From the Communications Act of 1934 as Amended

Section 1. For the purpose of regulating interstate and foreign commerce in communication by wire and radio as to make available, so far as possible, to all the people of the United States a rapid, efficient, nation-wide, and worldwide wire and communication service with adequate facilities at reasonable charges, for the purpose of the national defense, for the purpose of promoting safety of life and property through the use of wire and radio communication, and for the purpose of securing a more effective execution of this policy by centralizing authority heretofore granted by law to several agencies and by granting additional authority with respect to interstate and foreign commerce in wire and radio communication, there is hereby created a Commission to be known as the "Federal Communications Commission," which shall be constituted as hereinafter provided and which shall execute and enforce the provisions of this act.

Section 301. It is the purpose of this Act, among other things, to maintain the control of the United States over all the channels of interstate and foreign radio transmission; and to provide for the use of such channels, but not the ownership thereof, by persons for limited periods of time, under licenses granted by Federal authority, and no such license shall be construed to create any right, beyond the terms, conditions, and periods of the license. No person shall use or operate any apparatus for the transmission of energy or communications or signals by radio (a) from one place in any territory or possession of the United States or in the District of Columbia to another place in the same territory, possession, or district; or (b) from any state, territory, or possession of the United States, or from the District of Columbia to any other state, territory, or possession of the United States; or (c) from any place in any state, territory, or possession of the United States, or in the District of Columbia, to any place in any foreign country or to any vessel; or (d) within any state when the effects of such use extend beyond the borders of said state, or when interference is caused by such use or operation with the transmission of such energy, communications, or signals from within said state to any place beyond its borders, or from any place beyond its borders to any place within said state, or with the transmission or reception of such energy, communications, or signals from and/to places beyond the borders of said state; or (e) upon any vessel or aircraft of the United States; or (f) upon any other mobile stations within the jurisdiction of the United States, except under and in accordance with this Act and with a license in that behalf granted under the provisions of this Act.

Sec. 303. Except as otherwise provided in this Act, the Commissions from time to time, as public convenience, interest, or necessity requires, shall—

(m)(1) Have authority to suspend the license of any operator upon proof sufficient to satisfy the Commission that the licensee—

(a) Has violated any provision of this act, treaty, or convention binding on the United States which the Commission is authorized to administer, or any regulation made by the Commission under any such act, treaty, or convention; or

(c) Has willfully damaged or permitted radio apparatus or installations to be damaged; or

(d) Has transmitted superfluous radio communications or signals or communications containing profane or obscene words, language, or meaning, or has knowingly transmitted—

(1) False or deceptive signals or communications, or

(2) A call signal or letter which has not been assigned by proper authority to the station he is operating, or;

(e) Has willfully or maliciously interfered with any other radio communications or signals; or

(f) Has obtained or attempted to obtain, or has assisted another to obtain or attempt to obtain, an operator's license by fraudulent means.

(n) Have authority to inspect all radio installations associated with stations required to be licensed by any act or which are subject to the provisions of any act, treaty, or convention binding on the United States, to ascertain whether in construction, installation, and operation they conform to the requirements of the rules and regulations of the Commission, the provisions of any act, the terms of any treaty or convention binding on the United States, and the conditions of the license or other instrument of authorization under which they are constructed, installed, or operated.

Sec. 325 (a) (In part). No person within the jurisdiction of the United States shall knowingly utter or transmit, or cause to be uttered or transmitted, any false or fraudulent signal of distress.

Public Law No. 772, 80 Cong. 2nd Session reads as follows: 1464—Broadcasting obscene language. Whoever utters any obscene, indecent, or profane language by means of radio communication shall be fined not more than $10,000 or imprisoned not more than two years or both.

Sec. 502. Any person who willfully and knowingly violates any rule, regulation, restriction, or condition made or imposed by the Commission under authority of this act, or any rule, regulation, restriction, or condition made or imposed by any international radio or wire communications treaty or convention, or regulations annexed thereto, to which the United States is or may hereafter become a party, shall, in addition to any other penalties provided by law, be punished, upon conviction thereof, by a fine of not more than $500 for each and every day during which such offense occurs.

FCC Notice to Individuals Required by the Privacy Act of 1974

This notice is required by the Privacy Act of 1974, P.L. 93-579, December 31, 1974, 5 U.S.C. 552a (e) (3)

The solicitation of personal information requested in this application or report is authorized by the Communications Act of 1934.

The principal purpose(s) for which the information will be used in the case of an application is to determine if the benefit requested is consistent with the public interest, convenience, and necessity, or in the case of a report is to evaluate the data as required in the discharge of the regulatory responsibilities of the Commission.

The staff, consisting variously of attorneys, accountants, engineers, and applications examiners, will use the information to evaluate and render a judgement as to whether to grant or deny an application, or in the case of a report to evaluate the data for conformance to Commission rules and/or policies.

In the case of an application, if all the information requested is not provided, the application may be returned without action having been taken upon it or its processing may be delayed while a request is made to provide the missing information. Additionally, failure on the part of a licensee to submit a renewal application may result in the

expiration of the license. Furthermore, failure on
the part of a licensee of the Commission in the
Broadcast or Common Carrier Service to submit a
renewal application or a required report when due
may result in a monetary forfeiture being issued
against such licensee, or in the case of a Cable
Television system operator it may result in a cease
and desist proceeding. Therefore, extreme care
should be exercised in making certain that *all* necessary information is provided.

United States of America
Federal Communications Commission

Form Approved
GAO No. B-180227(R01 02)

FCC FORM 505

December 1974

APPLICATION FOR CLASS C OR D STATION
LICENSE IN THE CITIZENS RADIO SERVICE

Instructions

A. Use a typewriter or print clearly in capital letters. Stay within the boxes. Skip a box where a space would normally appear.
B. Sign and date application.
C. Enclose appropriate fee with application. DO NOT SUBMIT CASH. Make check or money order payable to Federal Communications Commission. No fee is required for an application filed by a governmental entity. For additional fee details, including amount and exemptions, see Subpart G of Part I, FCC Rules and Regulations.
D. Do not enclose order form or subscription fee for FCC Rules.

E. MAIL APPLICATION TO FEDERAL COMMUNICATIONS COMMISSION, GETTYSBURG, PA. 17325.

1. Complete if license is for an individual

Applicant's First Name   Init.   Last

2. Date of Birth

Month   Day   Year

3. Complete if license is for a business

Applicant's Name of Business, Organization, Or Partnership

4. Mailing Address (Number and Street) If P.O. Box or RFD# Is Used Also Fill Out Items 8—10.

5. City   6. State   7. Zip Code

NOTE:
Do not operate until you have your own license. Use of any call sign not your own is prohibited

8. If Item 4 is P.O. Box or RFD#, Give Address Or Location Of Principal Station

9. City   10. State

11. Type of Applicant (Check one)

☐ Individual   ☐ Association   ☐ Corporation
☐ Business Partnership   ☐ Governmental Entity
☐ Sole Proprietor or Individual/Doing Business As
☐ Other (Specify) _______________

12. This application is for

☐ New License
☐ Renewal
☐ Increase in Number of Transmitters

IMPORTANT
Give Current Call Sign

13. This application is for (Check only one)

☐ Class C Station License
(NON-VOICE—REMOTE CONTROL OF MODELS)

☐ Class D Station License (VOICE)

14. Indicate number of transmitters applicant will operate during the five year license period (Check one)

☐ 1 to 5   ☐ 6 to 15   ☐ 16 or more (Specify No. and attach statement justifying need.)

15.-Certification   I certify that:
• The applicant is not a foreign government or a representative thereof.
• The applicant has (or has ordered from the Government Printing Office) a current copy of Part 95 of the Commission's rules governing the Citizens Radio Service.
• The applicant will operate his transmitter in full compliance with the applicable law and current rules of the FCC and that his station will not be used for any purpose contrary to Federal, State, or local law or with greater power than authorized.
• The applicant waives any claim against the regulatory power of the United States relative to the use of a particular frequency or the use of the medium of transmission of radio waves because of any such previous use, whether licensed or unlicensed.

WILLFUL FALSE STATEMENTS MADE ON THIS FORM OR AT-TACHMENTS ARE PUNISHABLE BY FINE AND IMPRISONMENT. U.S. CODE, TITLE 18, SECTION 1001.

16. _______________
Signature of: Individual applicant, or authorized person on behalf of a governmental entity or partnership, or an officer of a corporation or association

17. Date _______________

---

remove at dotted line

ORDER FORM   Please Print Or Type

Please enter _______ subscription(s) to Volume VI, containing Parts 95, 97 and 99 of the Federal Communications Commission Rules and Regulations.

($5.35 per domestic subscription, which includes U.S. Territories and for Canada and Mexico; $6.70 per other foreign subscription.)

Name—First, Last

Company Name Or Additional Address Line

Street Address

City   State   Zip Code

☐ Remittance Enclosed (Make checks payable to Superintendent of Documents)
☐ Charge to my Deposit Account
No. _______________

MAIL ORDER FORM TO:
Superintendent of Documents
Government Printing Office
Washington, D.C. 20402

GPO : 1975 O — 568–584

RULES AND REGULATIONS

Part 95 | *Citizens Radio Service*

AMENDED 9-6-74

FEDERAL COMMUNICATIONS COMMISSION

CONTENTS

AUTHORITY: §§ 95.1 to 95.147 issued under secs. 4, 303, 48 Stat. 1066, 1082, as amended; 47 U.S.C. 154, 303. Interpret or apply 48 Stat. 1064-1068, 1081-1105, as amended; 47 U.S.C. Sub-chap. I, III-VI.

SUBPART A—GENERAL

§ 95.1 *Basis and purpose.* The rules and regulations set forth in this part are issued pursuant to the provisions of Title III of the Communications Act of 1934, as amended, which vests authority in the Federal Communications Commission to regulate radio transmissions and to issue licenses for radio stations. These rules are designed to provide for private short-distance radiocommunications service for the business or personal activities of licensees, for radio signaling, for the control of remote objects or devices by means of radio; all to the extent that these uses are not specifically prohibited in this part. They also provide for procedures whereby manufacturers of radio equipment to be used or operated in the Citizens Radio Service may obtain type acceptance and/or type approval of such equipment as may be appropriate.

§ 95.3 *Definitions.* For the purpose of this part, the following definitions shall be applicable. For other definitions, refer to Part 2 of this chapter.

(a) Definitions of services.

Citizens Radio Service. A radiocommunications service of fixed, land, and mobile stations intended for short-distance personal or business radiocommunications, radio signaling, and control of remote objects or devices by radio; all to the extent that these uses are not specifically prohibited in this part.

Fixed service. A service of radiocommunication between specified fixed points.

Mobile service. A service of radiocommunication between mobile and land stations or between mobile stations.

(b) Definitions of stations.

Base station. A land station in the land mobile service carrying on a service with land mobile stations.

Class A station. A station in the Citizens Radio Service licensed to be operated on an assigned frequency in the 460-470 MHz band with a transmitter output power of not more than 50 watts.

Class B station. (All operations terminated as of November 1, 1971.)

Class C station. A station in the Citizens Radio Service licensed to be operated on an authorized frequency in the 26.96-27.23 MHz band, or on the frequency 27.255 MHz, for the control of remote objects or devices by radio, or for the remote actuation of devices which are used solely as a means of attracting attention, or on an authorized frequency in the 72-76 MHz band for the radio control of models used for hobby purposes only.

Class D station. A station in the Citizens Radio Service licensed to be operated for radio-telephony, only, on an authorized frequency in the 26.96-27.23 MHz band and on the frequency 27.255 MHz.

Fixed station. A station in the fixed service.

Land station. A station in the mobile service not intended for operation while in motion. (Of the various types of land stations, only the base station is pertinent to this part.)

Mobile station. A station in the mobile service intended to be used while in motion or during halts at unspecified points. (For the purpose of this part, the term includes hand-carried and pack-carried units.)

(c) Miscellaneous definitions.

Antenna structures. The term "antenna structures" includes the radiating system, its supporting structures, and any appurtenances mounted thereon.

Assigned frequency. The frequency appearing on a station authorization from which the carrier frequency may deviate by an amount not to exceed that permitted by the frequency tolerance.

Authorized bandwidth. The maximum permissible bandwidth for the particular emission used. This shall be the occupied bandwidth, whichever is greater.

Carrier power. The average power at the output terminals of a transmitter (other than a transmitter having a suppressed, reduced or controlled carrier) during one radio frequency cycle under conditions of no modulation.

Control point. A control point is an operating position which is under the control and supervision of the licensee, at which a person immediately responsible for the proper operation of the transmitter is stationed, and at which adequate means are available to aurally monitor all transmissions and to render the transmitter inoperative.

Dispatch point. A dispatch point is any position from which messages may be transmitted under the supervision of the person at a control point.

Double sideband emission. An emission in which both upper and lower sidebands resulting from the modulation of a particular carrier are transmitted. The carrier or a portion thereof, also may be present in the emission.

Harmful interference. Any emission, radiation or induction which endangers the functioning of a radio-navigation service or other safety service or seriously degrades, obstructs or repeatedly interrupts a radio-communication service operating in accordance with applicable laws, treaties, and regulations.

Man-made structure. Any construction other than a tower, mast, or pole.

Mean power. The power at the output terminals of a transmitter during normal operation, averaged over a time sufficiently long compared with the period of the lowest frequency encountered in the modulation. A time of 1/10 second during which the mean power is greatest will be selected normally.

Necessary bandwidth. For a given class of emission, the minimum value of the occupied bandwidth sufficient to ensure the transmission of information at the rate and with the quality required for the system employed, under specified conditions. Emissions useful for the good functioning of the receiving equipment, as for example, the emission corresponding to the carrier of reduced carrier systems, shall be included in the necessary bandwidth.

Occupied bandwidth. The frequency bandwidth such that, below its lower and above its upper frequency limits, the mean powers radiated are each equal to 0.5% of the total mean power radiated by a given emission.

Omnidirectional antenna. An antenna designed so that the maximum radiation in any horizontal direction is within 3 dB of the minimum radiation in the horizontal direction.

Peak envelope power. The average power at the output terminals of a transmitter during one radio frequency cycle at the highest crest of the modulation envelope, taken under conditions of normal operation.

Person. The term "person" includes an individual, partnership, association, joint-stock company, trust, or corporation.

Remote control. The term "remote control" when applied to the use or operation of a citizens radio station means control of the transmitting equipment of that station from any place other than the location of the transmitting equipment, except that direct mechanical control or direct electrical control by wired connections of transmitting equipment from some other point on the same premises, craft, or vehicle shall not be considered to be remote control.

Single sideband emission. An emission in which only one sideband is transmitted. The carrier, or a portion thereof, also may be present in the emission.

Station authorization. Any construction permit, license, or special temporary authorization issued by the Commission.

[§ 95.3 (b) and (c) amended eff. 11-23-73; VI (72)-3]

§ 95.5 *Policy governing the assignment of frequencies.* (a) The frequencies which may be assigned to Class A stations in the Citizens Radio Service, and the frequencies which are available for use by Class C, or Class D stations are listed in Subpart C of this part. Each frequency available for assignment to, or use by, stations in this service is available on a shared basis only, and will not be assigned for the exclusive use of any one applicant; however, the use of a particular frequency may be restricted to (or in) one or more specified geographical areas.

(b) In no case will more than one frequency be assigned to Class A stations for the use of a single applicant in any given area until it has been demonstrated conclusively to the Commission that the assignment of an additional frequency is essential to the operation proposed.

(c) All applicants and licensees in this service shall cooperate in the selection and use of the frequencies assigned or

---

authorized, in order to minimize interferences and thereby obtain the most effective use of the authorized facilities.

(d) Simultaneous operation on more than one frequency in the 72-76 MHz band by a transmitter or transmitters of a single licensee is prohibited whenever such operation will cause harmful interference to the operation of other licensees in this service.

[§ 95.5 (a) amended eff. 12-7-73; VI (72)-3]

§ 95.6 *Types of operation authorized.* (a) Class A stations may be authorized as mobile stations, as base stations, as fixed stations, or as base or fixed stations to be operated at unspecified or temporary locations.

(b) Class C, and Class D stations are authorized as mobile stations only; however, they may be operated at fixed locations in accordance with other provisions of this part.

[§ 95.6 (b) ammended eff. 12-7-73; VI (72)-3]

§ 95.7 *General citizenship restrictions.* A station license may not be granted to or held by:

(a) Any alien or the representative of any alien;

(b) Any foreign government or the representative thereof;

(c) Any corporation organized under the laws of any foreign government;

(d) Any corporation of which any officer or director is an alien;

(e) Any corporation of which more than one-fifth of the capital stock is owned of record or voted by: Aliens or their representatives; a foreign government or representative thereof; or any corporation organized under the laws of a foreign country;

(f) Any corporation directly or indirectly controlled by any other corporation of which any officer or more than one-fourth of the directors are aliens, if the Commission finds that the public interest will be served by the refusal or revocation of such license; or

(g) Any corporation directly or indirectly controlled by any other corporation of which more than one-fourth of the capital stock is owned of record or voted by: Aliens or their representatives; a foreign government or representatives thereof; or any corporation organized under the laws of a foreign government, if the Commission finds that the public interest will be served by the refusal or revocation of such license.

SUBPART B—APPLICATIONS AND LICENSES

§ 95.11 *Station authorization required.* No radio station shall be operated in the Citizens Radio Service except under and in accordance with an authorization granted by the Federal Communications Commission.

§ 95.13 *Eligibility for station license.* (a) Subject to the general restrictions of §95.7, any person, other than an unincorporated association in the case of a Class D station, is eligible to hold an authorization to operate a station in the Citizens Radio Service: *Provided,* That if an applicant for a Class A or Class D station authorization is an individual or partnership, such individual or each partner is eighteen or more years of age; or if an applicant for a Class C station authorization is an individual or partnership, such individual or each partner is twelve or more years of age.

Note: While the basis of eligibility in this service includes any state, territorial, or local government entity, or any agency operating by the authority of such governmental entity, including any duly authorized state, territorial, or local civil defense agency, it should be noted that the frequencies available to stations in this service are shared without distinction between all licensees and that no protection is afforded to the communications of any station in this service from interference which may be caused by the authorized operation of other licensed stations.

(b) Notwithstanding the provisions of paragraph (a) of this section, an unincorporated association may be authorized to operate a Class D station in this service upon a showing satisfactory to the Commission that the proposed radio operations are not feasible, or may not be as efficient or economical, when conducted under station licenses issued to the individual members. A station license shall not be issued to an unincorporated association solely to avoid the operating restrictions on communications between stations licensed to different persons, contained elsewhere in this part. Unincorporated associations which hold Class D station licenses in this service as of November 1, 1964, must make the showing required by this paragraph upon application for renewal and/or modification of license. An unincorporated association, when licensed under the provisions of this paragraph, may upon specific prior approval of the Commission provide radiocommunications for its members.

(c) No person shall hold more than one Class C, and one Class D station license.

[§ 95.13 (a) and (c) amended eff. 12-7-73; VI (72)-3]

§ 95.15 *Filing of applications.* (a) To assure that necessary information is supplied in a consistent manner by all persons, standard forms are prescribed for use in connection with the majority of applications and reports submitted for Commission consideration. Standard numbered forms applicable to the Citizens Radio Service are discussed in § 95.19 and may be obtained from the Washington, D.C. 20554, office of the Commission, or from any of its engineering field offices.

(b) All formal applications for Class C, or Class D new, modified or renewal station authorizations shall be submitted to the Commission's office at 334 York Street, Gettysburg, Pa. 17325. Applications for Class A station authorizations, applications for consent to transfer of control of a corporation holding any citizens radio station authorization, requests for special temporary authority or other special requests, and correspondence relating to an application for any class citizens radio station authorization shall be submitted to the Commission's office at Washington, D.C. 20554. and should be directed to the attention of the Secretary. Beginning January 1, 1973, applicants for Class A stations in the Chicago Regional Area, defined in §95.19, shall submit their applications to the Commission's Regional Office. The address of the Regional Office will be announced at a later date. Applications involving Class A or Class D station equipment which is neither type approved nor crystal controlled, whether of commercial or home construction, shall be accompanied by supplemental data describing in detail the design and construction of the transmitter and methods employed in testing it to determine compliance with the technical requirements set forth in Subpart C of this part.

(c) Unless otherwise specified, an application shall be filed at least 60 days prior to the date on which it is desired that Commission action thereon be completed. In any case where the applicant has made timely and sufficient application for renewal of license, in accordance with the Commission's rules, no license with reference to any activity of a continuing nature shall expire until such application shall have been finally determined.

(d) Failure on the part of the applicant to provide all the information required by the application form, or to supply the necessary exhibits or supplementary statements may constitute a defect in the application.

(e) Applicants proposing to construct a radio station on a site located on land under the jurisdiction of the U.S. Forest Service, U.S. Department of Agriculture, or the Bureau of Land Management, U.S. Department of the Interior, must supply the information and must follow the procedure prescribed by § 1.70 of this chapter.

[§ 95.15 (b) amended and footnote 1 deleted eff. 12-7-73; VI (72)-3]

§ 95.17 *Who may sign applications.* (a) Except as provided in paragraph (b) of this section, applications, amendments thereto, and related statements of fact required by the Commission shall be personally signed by the applicant, if the applicant is an individual; by one of the partners, if the applicant is a partnership; by an officer, if the applicant is a corporation; or by a member who is an officer, if the applicant is an unincorporated association. Applications, amendments, and related statements of fact filed on behalf of eligible government entities, such as states and territories of the United States and political subdivisions thereof, the District of Columbia, and units of local government, including incorporated municipalities, shall be signed by such duly elected or appointed officials as may be competent to do so under the laws of applicable jurisdiction.

(b) Applications, amendments thereto, and related statements of fact required by the Commission may be signed by the applicant's attorney in case of the applicant's physical disability or of his absence from the United States. The attorney shall in that event separately set forth the reason why the application is not signed by the applicant. In addition, if any matter is slated on the basis of the attorney's belief only (rather than his knowledge), he shall separately set forth his reasons for believing that such statements are true.

(c) Only the original of applications, amendments, or related statements of fact need be signed; copies may be conformed.

(d) Applications, amendments, and related statements of fact need not be signed under oath. Willful false statements made therein, however, are punishable by fine and imprisonment, U.S. Code, Title 18, section 1001, and by appropriate administrative sanctions, including revocation of station license pursuant to section 312(a) (1) of the Communications Act of 1934, as amended.

§ 95.19 *Standard forms to be used.* (a) *FCC Form 505, Application for Class C or D Station License in the Citizens Radio Service.* This form shall be used when:

(1) Application is made for a new Class C, or Class D authorization. A separate application shall be submitted for each proposed class of station.

(2) Application is made for modification of any existing Class C or Class D station authorization those cases where prior Commission approval of certain changes is required (see § 95.35).

(3) Application is made for renewal of an existing Class C or Class D station authorization, or for reinstatement of such an expired authorization.

[§ 95.19 (a) and (a) (3) amended eff. 12-7-73; VI (72)-3]

(b) *FCC Form 400, Application for Radio Station Authorization in the Safety and Special Radio Services.* Except as provided in paragraph (d) of this section, this form shall be used when:

(1) Application is made for a new Class A base station or fixed station authorization. Separate applications shall be submitted for each proposed base or fixed station at different fixed locations; however, all equipment intended to be operated at a single fixed location is considered to be one station which may, if necessary, be classed as both a base station and a fixed station.

(2) Application is made for a new Class A station authorization for any required number of mobile units (including hand-carried and pack-carried units) to be operated as a group in a single radiocommunication system in a particular area. An application for Class A mobile station authorization may be combined with the application for a single Class A base station authorization when such mobile units are to be operated with that base station only.

(3) Application is made for station license of any Class A base station or fixed station upon completion of construction or installation in accordance with the terms and conditions set forth in any construction permit required to be issued for that station, or application for extension of time within which to construct such a station.

(4) Application is made for modification of any existing Class A station authorization in those cases where prior Commission approval of certain changes is required (see § 95.35).

(5) Application is made for renewal of an existing Class A station authorization, or for reinstatement of such an expired authorization.

(6) [Reserved]

(7) Application is made for an authorization for a new Class A base or fixed station to be operated at unspecified or temporary locations. When one or more individual transmitters are each intended to be operated as a base station or as a fixed station at unspecified or temporary locations for indeterminate periods, such transmitters may be considered to comprise a single station intended to be operated at temporary locations. The application shall specify the general geographic area within which the operation will be confined. Sufficient data must be submitted to show the need for the proposed area of operation.

(c) *FCC Form 703, Application for Consent to Transfer of Control of Corporation Holding Construction Permit or Station License.* This form shall be used when application is made for consent to transfer control of a corporation holding any citizens radio station authorization.

(d) Beginning April 1, 1972, FCC Form 425 shall be used in lieu of FCC Form 400.

§ 95.25 *Amendment or dismissal of application.* (a) Any application may be amended upon request of the applicant as a matter of right prior to the time the application is granted or designated for hearing. Each amendment to an application shall be signed and submitted in the same manner and with the same number of copies as required for the original application.

(b) Any application may, upon written request signed by the applicant or his attorney, be dismissed without prejudice as a matter of right prior to the time the application is granted or designated for hearing.

§ 95.27 *Transfer of license prohibited.* A station authorization in the Citizens Radio Service may not be transferred or assigned. In lieu of such transfer or assignment, an application for new station authorization shall be filed in each case, and the previous authorization shall be forwarded to the Commission for cancellation.

§ 95.29 *Defective applications.* (a) If an applicant is requested by the Commission to file any documents or information not included in the prescribed application form, a failure to comply with such requests will constitute a defect in the application.

(b) When an application is considered to be incomplete or defective, such application will be returned to the applicant, unless the Commission may otherwise direct. The reason for return of the applications will be indicated, and if appropriate, necessary additions or corrections will be suggested.

§ 95.31 *Partial grant.* Where the Commission, without a hearing, grants an application in part, or with any privileges, terms, or conditions other than those requested, the action of the Commission shall be considered as a grant of such application unless the applicant shall, within 30 days from the date on which such grant is made, or from its effective date of later date is specified, file with the Commission, a written rejection of the grant as made. Upon receipt of such rejection, the Commission will vacate its original action upon the application and, if appropriate, set the application for hearing.

§ 95.33 *License term.* Licenses for stations in the Citizens Radio Service will normally be issued for a term of 5 years from the date of original issuance, major modification, or renewal.

[§95.33 amended eff. 9-15-70; VI(69)-5]

§ 95.35 *Changes in transmitters and authorized stations.* Authority for certain changes in transmitters and authorized stations must be obtained from the Commission before the changes are made, while other changes do not require prior Commission approval. The following paragraphs of this section describe the conditions under which prior Commission approval is or is not necessary.

(a) Proposed changes which will result in operation inconsistent with any of the terms of the application for modification of license be submitted to the Commission. Application for modification shall be submitted in the same manner as an application for a new station license, and the licensee shall forward his existing authorization to the Commission for cancellation immediately upon receipt of the superseding authorization. Any of the following changes to authorized stations may be made only upon approval by the Commission:

(1) Increase the overall number of transmitters authorized.

(2) Change the presently authorized location of a Class A fixed or base station or control point.

(3) Move, change the height of, or erect a Class A station antenna structure.

(4) Make any change in the type of emission or any increase in bandwidth of emission or power of a Class A station.

(5) Addition or deletion of control point(s) for an authorized transmitter of a Class A station.

(6) Change or increase the area of operation of a Class A mobile station or a Class A base or fixed station authorized to be operated at temporary locations.

(7) Change the operating frequency of a Class A station.

(b) When the name of a licensee is changed (without changes in the ownership, control, or corporate structure), or when the mailing address of the licensee is changed (without changing the authorized location of the base or fixed Class A station) a formal application for modification of the license is not required. However, the licensee shall notify the Commission promptly of these changes. The notice, which may be in letter form, shall contain the name and address of the licensee as they appear in the Commission's records, the new name and/or address, as the case may be, and the call signs and classes of all radio stations authorized to the licensee under this part. The notice concerning Class C or D radio stations shall be sent to Federal Communications Commission, Gettysburg, Pa., 17325, and a copy shall be maintained with the records of the station. The notice concerning Class A stations shall be sent to (1) Secretary, Federal Communications Commission, Washington, D.C., 20554, and (2) to Engineer in Charge of the Radio District in which the station is located, and a copy shall be maintained with the license of the station until a new license is issued.

(c) Proposed changes which will not depart from any of the terms of the outstanding authorization for the station may be made without prior Commission approval. Included in such changes is the substitution of transmitting equipment at any station, provided that the equipment employed is included in the Commission's "Radio Equipment List," and is listed as acceptable for use in the appropriate class of station in this service. Provided it is crystal-controlled and otherwise complies with the power, frequency tolerance, emission and modulation percentage limitations prescribed, non-type accepted equipment may be substituted at:

(1) Class C stations operated on frequencies in the 26.99-27.26 MHz band;

(2) Class D stations until November 22, 1974.

(d) Transmitting equipment type accepted for use in Class D stations shall not be modified by the user. Changes which are specifically prohibited include:

(1) Internal or external connection or addition of any part, device or accessory not included by the manufacturer with the transmitter for its type acceptance. This shall not prohibit the external connection of antennas or antenna transmission lines, antenna switches, passive networks for coupling transmission lines or antennas to transmitters, or replacement of microphones.

(2) Modification in any way not specified by the transmitter manufacturer and not approved by the Commission.

(3) Replacement of any transmitter part by a part having different electrical characteristics and ratings from that replaced unless such part is specified as a replacement by the transmitter manufacturer.

(4) Substitution or addition of any transmitter oscillator crystal unless the crystal manufacturer or transmitter manu-

facturer has made an express determination that the crystal type, as installed in the specific transmitter type, will provide that transmitter type with the capability of operating within the frequency tolerance specified in Section 95.45 (a).

(5) Addition or substitution of any component, crystal or combination of crystals, or any other alteration to enable transmissions on any frequency not authorized for use by the licensee.

(e) Only the manufacturer of the particular unit of equipment type accepted for use in Class D stations may make the permissive changes allowed under the provisions of Part 2 of this chapter for type acceptance. However, the manufacturer shall not make any of the following changes to the transmitter without prior authorization from the Commission:

(1) Addition of any accessory or device not specified in the application for type acceptance and approved by the Commission in granting said type acceptance.

(2) Addition of any switch, control, or external connection.

(3) Modification to provide capability for an additional number of transmitting frequencies.

[§ 95.35 headnote and intro. text revised; (c) amended, (d) and (e) added eff. 11-23-73; (b) amended eff. 12-7-73; VI (72)-3]

§ 95.37 *Limitations on antenna structures.*

(a) Except as provided in paragraph (b) of this section, an antenna for a Class A station which exceeds the following height limitations may not be erected or used unless notice has been filed with both the FAA on FAA Form 7460-1 and with the Commission on Form 714 or on the license application form, and prior approval by the Commission has been obtained for:

(1) Any construction or alteration of more than 200 feet in height above ground level at its site (§ 17.7(a) of this chapter).

(2) Any construction or alteration of greater height than an imaginary surface extending outward and upward at one of the following slopes (§ 17.7(b) of this chapter):

(i) 100 to 1 for a horizontal distance of 20,000 feet from the nearest point of the nearest runway of each airport with at least one runway more than 3,200 feet in length, excluding heliports, and seaplane bases without specified boundaries, if that airport is either listed in the Airport Directory of the current Airman's Information Manual or is operated by a Federal military agency.

(ii) 50 to 1 for a horizontal distance of 10,000 feet from the nearest point of the nearest runaway of each airport with its longest runway no more than 3,200 feet in length, excluding heliports, and seaplane bases without specified boundaries, if that airport is either listed in the Airport Directory or is operated by a Federal military agency.

(iii) 25 to 1 for a horizontal distance of 5,000 feet from the nearest point of the nearest landing and takeoff area of each heliport listed in the Airport Directory or operated by a Federal military agency.

(3) Any constructon or alteration on any airport listed in the Airport Directory of the current Airman's Information Manual (§ 17.7(c) of this chapter).

(b) A notification to the Federal Aviation Administration is not required for any of the following construction or alteration of Class A station antenna structures.

(1) Any object that would be shielded by existing structures of a permanent and substantial character or by natural terrain or topographic features of equal or greater height, and would be located in the congested area of a city, town, or settlement where it is evident beyond all reasonable doubt that the structure so shielded will not adversely affect safety in air navigation. Applicants claiming such exemption shall submit a statement with their application to the Commission explaining the basis in detail for their finding (§ 17.14 (a) of this chapter).

(2) Any antenna structure of 20 feet or less in height except one that would increase the height of another antenna structure (§ 17.14(b) of this chapter).

(c) A Class C or Class D station operated at a fixed location shall employ a transmitting antenna which complies with at least one of the following:

(1) The antenna and its supporting structure does not exceed 20 feet in height above ground level; or

(2) The antenna and its supporting structure does not exceed by more than 20 feet the height of any natural formation, tree or manmade structure on which it is mounted; or

NOTE: A manmade structure is any construction other than a tower, mast, or pole.

(3) The antenna is mounted on the transmitting antenna structure of another authorized radio station and does not exceed the height of the antenna supporting structure of the other station; or

(4) The antenna is mounted on and does not exceed the height of an antenna structure otherwise used solely for receiving purposes, which structure itself complies with subparagraph (1) or (2) of

this paragraph.

(5) The antenna is omnidirectional and the highest point of the antenna or its supporting structure does not exceed 60 feet above ground level and the highest point also does not exceed 1 foot in height above the established airport elevation for each 100 feet of horizontal distance from the nearest point of the nearest airport runway.

NOTE: A worksheet will be made available upon request to assist in determining the maximum permissible height of an antenna structure.

(d) Class C stations operated on frequencies in the 72-76 MHz band shall employ a transmitting antenna which complies with all of the following:

(1) The gain of the antenna shall not exceed that of a half-wave dipole;

(2) The antenna shall be immediately attached to, and an integral part of, the transmitter; and

(3) Only vertical polarization shall be used.

(e) Further details as to whether an aeronautical study and/or obstruction marking and lighting may be required, and specifications for obstruction marking and lighting when required, may be obtained from Part 17 of this chapter, "Construction, Marking and Lighting of Antenna Structures."

[§ 95.37 (c) amended eff. 12-7-73; VI (72)-3]

SUBPART C—TECHNICAL REGULATIONS

§ 95.41 *Frequencies available.* (a) Frequencies available for assignment to Class A stations:

(1) The following frequencies or frequency pairs are available primarily for assignment to base and mobile stations. They may also be assigned to fixed stations as follows:

(i) Fixed stations which are used to control base stations of a system may be assigned the frequency assigned to the mobile units associated with the base station. Such fixed stations shall comply with the following requirements if they are located within 75 miles of the center of urbanzied areas of 200,000 or more population.

(a) If the station is used to control one or more base stations located within 45 degrees of azimuth, a directional antenna having a front-to-back ratio of at least 15 dB shall be used at the fixed station. For other situations where such a directional antenna cannot be used, a cardioid, bidirectional or omnidirectional antenna may be employed. Consistent with reasonable design, the antenna used must, in each case, produce a radiation pattern that provides only the coverage necessary to permit satisfactory control of each base station and limit radiation in other directions to the extent feasible.

(b) [Reserved]

[§ 95.41 (b) deleted eff. 12-7-73; VI (72)-3]

(c) Each application for a control station to be authorized under the provisions of this paragraph shall be accompanied by a statement certifying that the output power of the proposed station transmitter will be adjusted to comply with the foregoing signal limitation. Records of the measurements used to determine the signal ratio shall be kept with the station records and shall be made available for inspection by Commission personnel upon request.

(d) Urbanized areas of 200,000 or more population are defined in the U.S. Census of Population, 1960, Vol. 1, table 23, page 50. The centers of urbanized areas are determined from the Appendix, page 226 of the U.S. Commerce publication "Air Line Distance Between Cities in the United States."

(ii) Fixed stations, other than those used to control base stations, which are located 75 or more miles from the center of an urbanized area of 200,000 or more population are listed on page 226 of the Appendix to the U.S. Department of Commerce publication "Air Line Distance Between Cities in the United States." When the fixed station is located 100 miles or less from the center of such an urbanized area, the power output may not exceed 15 watts. All fixed systems are limited to a maximum of two frequencies and must employ directional antennas with a front-to-back ratio of at least 15 dB. For two-frequency systems, separation between transmit-receive frequencies is 5 MHz.

Base and Mobile (MHz)	Mobile only (MHz)
462.550	467.550
462.575	467.575
462.600	467.600
462.625	467.625
462.650	467.650
462.675	467.675
462.700	467.700
462.725	467.725

(2) Conditions governing the operation of stations authorized prior to March 18, 1968:

(i) All base and mobile stations authorized to operate on frequencies other than those listed in subparagraph (1) of this paragraph may continue to operate on those frequencies only until January 1, 1970.

(ii) Fixed stations located 100 or more miles from the center of any urbanized area of 200,000 or more population authorized to operate on frequencies other than those listed in subparagraph (1) of this paragraph will not have to change frequencies provided no interference is caused to the operation of stations in the land mobile service.

(iii) Fixed stations, other than those used to control base stations, located less than 100 miles (75 miles if the transmitter power output does not exceed 15 watts) from the center of any urbanized area of 200,000 or more population must discontinue operation by November 1, 1971. However, any operation after January 1, 1970, must be on frequencies listed in subparagraph (1) of this paragraph.

(iv) Fixed stations located less than 100 miles from the center of any urbanized area of 200,000 or more population, which are used to control base stations and are authorized to operate on frequencies other than those listed in subparagraph (1) of this paragraph may continue to operate on those frequencies only until January 1, 1970.

(v) All fixed stations must comply with the applicable technical requirements of subparagraph (1) relating to antennas and radiated signal strength of this paragarph by November 1, 1971.

(vi) Notwithstanding the provisions of subdivision (i) through (v) of this subparagraph, all stations authorized to operate on frequencies between 465.000 and 465.500 MHz and located within 75 miles of the center of the 20 largest urbanized areas of the United States, may continue to operate on these frequencies only until January 1, 1969. An extension to continue operation on such frequencies until January 1, 1970, may be granted to such station licensees on a case by case basis if the Commission finds that continued operation would not be inconsistent with planned usage of the particular frequency for police purposes. The 20 largest urbanized areas can be found in the U.S. Census of Population, 1960, vol. 1, table 23, page 50. The centers of urbanized areas are determined from the appendix, page 226, of the U.S. Commerce publication, "Air Line Distance Between Cities in the United States."

(b) Effective March 18, 1968, frequencies will no longer be available to Class B stations for new or modified facilities. However, existing Class B stations may upon renewal, continue to operate until November 1, 1971.

(c) Class C mobile stations may employ only amplitude tone modulation or on-off keying of the unmodulated carrier, on a shared basis with other stations in the Citizens Radio Service on the frequencies and under the conditions specified in the following tables:

(1) For the control of remote objects or devices by radio, or for the remote actuation of devices which are used solely as a means of attracting attention and subject to no protection from interference due to the operation of industrial, scientific, or medical devices within the 26.96-27.28 MHz band, the following frequencies are available:

(MHz)	(MHz)	(MHz)
26.995	27.095	27.195
27.045	27.145	¹27.255

¹The frequency 27.255 MHz also is shared with stations in other services.

(2) Subject to the conditions that interference will not be caused to the remote control of industrial equipment operating on the same or adjacent frequencies and to the reception of television transmissions on Channels 4 or 5; and that no protection will be afforded from interference due to the operation of fixed and mobile stations in other services assigned to the same or adjacent frequencies in the band, the following frequencies are available solely for the radio remote control of models used for hobby purposes:

(i) For the radio remote control of any model used for hobby purposes:

(MHz)	(MHz)	(MHz)
72.16	72.32	72.96

(ii) For the radio remote control of aircraft models only:

(MHz)	(MHz)	(MHz)
72.08	72.24	72.40
75.64		

(d) The frequencies listed in the following tables are available for use by Class D mobile stations employing radiotelephony only, on a shared basis with other stations in the Citizens Radio Service, and subject to no protection from interference due to the operation of industrial, scientific, or medical devices within the 26.96-27.28 MHz band.

(1) The following frequencies, commonly known as Channels 1 through 8 and 10 through 23, may be used for communications between units of the same station:

(MHz)	Channel	(MHz)	Channel
26.965	1	27.115	13
26.975	2	27.125	14
26.985	3	27.135	15
27.005	4	27.155	16
27.015	5	27.165	17
27.025	6	27.175	18
27.035	7	27.185	19
27.055	8	27.205	20
27.075	10	27.215	21
27.085	11	27.225	22
27.105	12	27.255	23

(2) Only the following frequencies may be used for communications between units of different stations:

(MHz)	Channel	(MHz)	Channel
27.075	10	27.125	14
72.085	11	27.135	15
27.105	12	27.255	23
27.115	13		

(3) The frequency 27.065 MHz (Channel 9) shall be used solely for:

(i) Emergency communications involving the immediate safety of life of individuals or the immediate protection of property or

(ii) Communications necessary to render assistance to a motorist.

NOTE: A licensee, before using Channel 9, must make a determination that his communication is either or both (a) an emergency communication or (b) is necessary to render assistance to a motorist. To be an emergency communication, the message must have some direct relation to the immediate safety of life or immediate protection of property. If no immediate action is required, it is not an emergency. What may not be an emergency under one set of circumstances may be an emergency under different circumstances. There are many worthwhile public service communications that do not qualify as emergency communications. In the case of motorist assistance, the message must be necessary to assist a particular motorist and not, except in a valid emergency, motorists in general. If the communications are to be lengthy, the exchange should be shifted to another channel, if feasible, after contact is established. No nonemergency or nonmotorist assistance communications are permitted on Channel 9 even for the limited purpose of calling a licensee monitoring a channel to ask him to switch to another channel. Although Channel 9 may be used for marine emergencies, it should not be considered a substitute for the authorized marine distress system. The Coast Guard has stated it will not "participate directly in the Citizens Radio Service by fitting with and/or providing a watch on any Citizens Band Channel. (Coast Guard Commandant Instruction 2302.6)"

The following are examples of permitted and prohibited types of communications. They are guidelines and are not intended to be all inclusive.

Permitted	Example message
Yes	"A tornado sighted six miles north of town."
No	"This is observation post number 10. No tornados sighted."
Yes	"I am out of gas on Interstate 95."
No	"I am out of gas in my driveway."
Yes	"There is a four-car collision at Exit 10 on the Beltway, send police and ambulance."
No	"Traffic is moving smoothly on the Beltway."
Yes	"Base to Unit 1, the Weather Bureau has just issued a thunderstorm warning. Bring the sailboat into port."
No	"Attention all motorists. The Weather Bureau advises that the snow tomorrow will accumulate 4 to 6 inches."
Yes	"There is a fire in the building on the corner of 6th and Main Streets."
No	"This is Halloween patrol unit number 3. Everything is quiet here."

The following priorities should be observed in the use of Channel 9.

1. Communications relating to an existing situation dangerous to life or property, i.e., fire, automobile accident.

2. Communications relating to a potentially hazardous situation, i.e., car stalled in a dangerous place, lost child, boat out of gas.

3. Road assistance to a disabled vehicle on the highway or street.

4. Road and street directions.

(e) Upon specific request accompanying application for renewal of station authorization, a Class A station in this service, which was authorized to operate on a frequency in the 460-461 MHz band until March 31, 1967, may be assigned that frequency for continued use until not later than March 31, 1968, subject to all other provisions of this part.

§ 95.43 *Transmitter power.* (a) Transmitter power is the power at the transmitter output terminals and delivered to the antenna, antenna transmission line, or and other impedance-matched, radio frequency load.

(1) For single sideband transmitters and other transmitters employing a reduced carrier, a suppressed carrier or a controlled carrier, used at Class D stations, transmitter power is the peak envelope power.

(2) For all transmitters other than those covered by paragraph (a) (1) of this section, the transmitter power is the carrier power.

(b) The transmitter power of a station shall not exceed the following values under any condition of modulation or other circumstances.

Class of station:	Transmitter power in watts
A	50
C—27.255 MHz	25
C—26.995-27.195 MHz	4
C—72-76 MHz	0.75
D—Carrier (where applicable)	4
D—Peak envelope power (where applicable)	12

[§ 95.43 headnote and text revised eff. 11-23-73; VI (72)-3]

§ 95.45 *Frequency tolerance.* (a) except in paragraphs (b) and (c) of this section, the carrier frequency of a transmitter in this service shall be maintained within the following percentage of the authorized frequency:

Class of station	Frequency Tolerance	
	Fixed and base	Mobile
A	0.00025	0.0005
C		.005
D		.005

(b) Transmitters used as Class C stations operating on authorized frequencies between 26.99 and 27.26 MHz with 2.5 watts or less mean output power, which are used solely for the control of remote objects or devices by radio (other than devices used solely as a means of attracting attention), are permitted a frequency tolerance of 0.01 percent.

(c) Class A stations operated at a fixed location used to control base stations, through use of a mobile only frequency, may operate with a frequency tolerance of 0.0005 percent.

[§ 95.45 amended eff. 11-23-73; VI (72)-3]

§ 95.47 *Types of emission.* (a) Except as provided in paragraph (e) of this section, Class A stations in this service will normally be authorized to transmit radiotelephony only. However, the use of tone signals or signaling devices solely to actuate receiver circuits, such as tone operated squelch or selective calling circuits, the primary function of which is to establish or establish and maintain voice communications, is permitted. The use of tone signals solely to attract attention is prohibited.

(b) [Reserved]

(c) Class C stations in this service are authorized to use amplitude tone modulation or on-off unmodulated carrier only, for the control of remote objects or devices by radio or for the remote actuation of devices which are used solely as a means of attracting attention. The transmission of any form of telegraphy, telephony or record communications by a Class C station is prohibited. Telemetering, except for the transmission of simple, short duration signals indicating the presence or absence of a condition or the occurrence of an event, is also prohibited.

(d) Transmitters used as Class D stations in this service are authorized to use amplitude voice modulation, either single or double sideband. Tone signals or signalling devices may be used only to actuate receiver circuits, such as tone operated squelch or selective calling circuits, the primary function of which is to establish or maintain voice communications. The use of any signals solely to attract attention or for the control of remote objects or devices is prohibited.

(e) Other types of emission not described in paragraph (a) of this section may be authorized for Class A citizens radio stations upon a showing of need therefor. An application requesting such authorization shall fully describe the emission desired, shall indicate the bandwidth required for satisfactory communication, and shall state the purpose for which such emission is required. For information regarding the classification of emissions and the calculation of bandwidth, reference should be made to Part 2 of this chapter.

[§ 95.47 (d) amended eff. 11-23-73; (b) deleted eff. 12-7-73; VI (72)-3]

§ 95.49 *Emission limitations.* (a) Each authoriztion issued to a Class A citizens radio station will show, as a prefix to the classification of the authorized emission, a figure specifying the maximum bandwidth to be occupied by the emission.

(b) [Reserved]

(c) The authorized bandwidth of the emission of any transmitter employing ampltiude modulation shall be 8 kHz for double sideband, 4 kHz for single sideband and the authorized bandwidth of the emission of transmitters employing frequency of phase modulation (Class F2 or F3 shall be 20 kHz. The use of Class F2 and F3 emissions in the frequency band 26.96-27.28 MHz is not authorized.

(d) The mean power of emissions shall be attenuated below the mean power of the transmitter in accordance with the following schedule:

(1) When using emissions other than single sideband:

(i) On any frequency removed from the center of the authorized bandwidth by more than 50 percent up to and including 100 percent of the authorized bandwidth: at least 25 decibels;

(ii) On any frequency removed from the center of the authorized bandwidth by more than 100 percent up to and including 250 percent of the authorized bandwidth: At least 35 decibels;

(2) When using single sideband emissions:

(i) On any frequency removed from the center of the authorized bandwidth by more than 50 percent up to and including 150 percent of the authorized bandwidth: At least 25 decibels:

(ii) On any frequency removed from the center of the authorized bandwidth by more than 150 percent up to and including 250 percent of the authorized bandwidth: At least 35 decibels;

(3) On any frequency removed from the center of the authorized bandwidth: At least 43 plus 10 $\log_{10}$ (mean power in watts) decibels.

(e) When an unauthorized emission results in harmful interference, the Commission may, in its discretion, require appropriate technical changes in equipment to alleviate the interference.

[§ 95.49 (c) and (d) amended eff. 11-23-73; (b) deleted eff. 12-7-73; VI (72)-31]

§ 95.51 *Modulation requirements.* (a) When double sideband, amplitude modulation is used for telephony the, modulation percentage shall be sufficient to provide efficient communication and shall not exceed 100 percent.

(b) Each transmitter for use in Class D stations, other than single sideband, suppressed carrier, or controlled carrier, for which type acceptance is requested after May 4, 1974, having more than 2.5 watts maximum output power shall be equipped with a device which automatically prevents modulation in excess of 100 percent on positive and negative peaks.

(c) The maximum audio frequency required for satisfactory radiotelephone intelligibility for use in this service is considered to be 3000 Hz.

(d) Transmitters for use at Class A stations shall be provided with a device which automatically will prevent greater than normal audio level from causing modulation in excess of that specified in this subpart; Provided however, That the requirements of this paragraph shall not apply to transmitters authorized at mobile stations and having an output power of 2.5 watts or less.

(e) Each transmitter of a Class A station which is equipped with a modulation limiter in accordance with the provisions of paragraph (d) of this section shall also be equipped with an audio low-pass filter. This audio low-pass filter shall be installed between the modulation limiter and the modulated stage and, at audo frequencies between 3 kHz and 20 kHz, shall have an attenuation greater than the attenuation at 1 kHz by at least:

60 $\log_{10}$ (f/3) decibels

where "f" is the audio frequency in kHz. At audio frequencies above 20 kHz, the attenuation shall be at least 50 decibels greater than the attenuation at 1 kHz.

(f) Simultaneous amplitude modulation and frequency or phase modulation of a transmitter is not authorized.

(g) The maximum frequency deviation of frequency modulated transmitters used at Class A stations shall not exceed ±5 kHz.

[§ 95.51 amended eff. 11-23-73; VI (72)-31]

§ 95.53 *Compliance with technical requirements.* (a) Upon receipt of notification from the Commission of a deviation from the technical requirements of the rules in this part, the radiations of the transmitter involved shall be suspended immediately, except for necessary tests and adjustments, and shall not be resumed until such deviation has been corrected.

(b) When any citizens radio station licensee receives a notice of violation indicating that the station has been operated contrary to any of the provisions contained in Subpart C of this part, or where it otherwise appears that operation of a station in this service may not be in accordance with applicable technical standards, the Commission may require the licensee to conduct such tests as may be necessary to determine whether the equipment is capable of meeting these standards and to make such adjustments as may be necessary to assure compliance therewith. A licensee who is notified that he is required to conduct such tests and/or make adjustments must, within the time limit specified in the notice, report to the Commission the results thereof.

(c) All tests and adjustments which may be required in accordance with paragraph (b) of this section shall be made by, or under the immediate supervision of, a person holding a first- or second-class commercial operator license, either radiotelephone or radio telegraph as may be appropriate for the type of emission employed. In each case, the report which is submitted to the Commission shall be signed by the licensed commercial operator. Such report shall describe the results of the tests and adjustments, the test equipment and procedures used, and shall state the type, class, and serial number of the operator's license. A copy of this report shall also be kept with the station records.

§ 95.55 *Acceptability of transmitters for licensing.* Transmitters type approved or type accepted for use under this part are included in the Commission's Radio Equipment List. Copies of this list are available for public reference at the Commission's Washington, D.C., offices and field offices. The requirements for transmitters which may be operated under a license in this service are set forth in the following paragraphs.

(a) Class A stations: All transmitters shall be type accepted.

(b) Class C stations:

(1) Transmitters operated in the band 72-76 MHz shall be type accepted.

(2) All transmitters operated in the band 26.99-27.26 MHz shall be type approved, type accepted or crystal controlled.

(c) Class D Stations:

(1) All transmitters first licensed, or marketed as specified in § 2.805 of this chapter, prior to November 22, 1974, shall be type accepted or crystal controlled.

(2) All transmitters first licensed, or marketed as specified in § 2.803 of this chapter, on or after November 22, 1974, shall be type accepted.

(3) Effective November 23, 1978, all transmitters shall be type accepted.

(4) Transmitters which are equipped to operate on any frequency not included in § 95.41 (d) (1) may not be installed at, or used by, any Class D station unless there is a station license posted at the transmitter location, or a transmitter identification card (FCC Form 452-C) attached to the transmitter, which indicates that operation of the transmitter on such frequency has been authorized by the Commission.

(d) With the exception of equipment type approved for use at a Class C station, all transmitting equipment authorized in this service shall be crystal controlled.

(e) No controls, switches or other functions which can cause operation in violation of the technical regulations of this part shall be accessible from the operating panel or exterior to the cabinet enclosing a transmitter authorized in this service.

§ 95.57 *Procedure for type acceptance of equipment.* (a) Any manufacturer of a transmitter built for use in this service, except noncrystal controlled transmitters for use at Class C stations, may request type acceptance for such transmitter in accordance with the type acceptance requirements of this part, following the type acceptance procedure set forth in Part 2 of this chapter.

(b) Type acceptance for an individual transmitter may also be requested by an applicant for a station authorization by following the type acceptance procedures set forth in Part 2 of this chapter. Such transmitters, if accepted, will not normally be included on the Commission's "Radio Equipment List", but will be individually enumerated on the station authorization.

(c) Additional rules with respect to type acceptance are set forth in Part 2 of this chapter. These rules include information with respect to withdrawal of type acceptance, modification of type-accepted equipment, and limitations on the findings upon which type acceptance is based.

(d) Transmitters equipped with a frequency or frequencies not listed in § 95.41 (d) (1) will not be type accepted for use at Class D stations unless the transmitter is also type accepted for use in the service in which the frequency is authorized, if type acceptance in that service is required.

§ 95.58 *Additional requirements for type acceptance.* (a) All transmitters shall be crystal controlled.

(b) Except for transmitters type accepted for use at Class A stations, transmitters shall not include any provisions for increasing power to levels in excess of the pertinent limits specified in Section 95.43.

(c) In addition to all other applicable technical requirements set forth in this part, transmitters for which type acceptance is requested after May 24, 1974, for use at Class D stations shall comply will the following:

(1) Single sideband transmitters and other transmitters employing reduced, suppressed or controlled carrier shall include a means of automatically preventing the transmitter power from exceeding either the maximum permissible peak envelope power or the rated peak envelope power of the transmitter, whichever is lower.

(2) Multi-frequency transmitters shall not provide more than 23 transmitting frequencies, and the frequency selector shall be limited to a single control.

(3) Other than the channel selector switch, all transmitting frequency determining circuitry, including crystals, employed in Class D station equipment shall be internal to the equipment and shall not be accessible from the exterior of the equipment cabinet or operating panel.

(4) Single sideband transmitters shall be capable of transmitting on the upper sideband. Capability for transmission also on the lower sideband is permissible.

(5) The total dissipation ratings, established by the manufacturer of the electron tubes or semiconductors which supply radio frequency power to the antenna terminals of the transmitter, shall not exceed 10 watts. For electron tubes, the rating shall be the Intermittent Commercial and Amateur Service (ICAS plate dissipation value if established.) For semiconductors, the rating shall be the collector or device dissipation value, whichever is greater, which may be temperature de-rated to not more than 50°C.

(d) Only the following external transmitter controls, connections or devices will normally be permitted in transmitters for which type acceptance is requested after May 24, 1974, for use at Class D stations. Approval of additional controls, connections or devices may be given after consideration of the function to be performed by such additions.

(1) Primary power connections. (Circuitry or devices such as rectifiers, transformers, or inverters which provide the nominal rated transmitter primary supply voltage may be used without voiding the transmitter type acceptance.)

(2) Microphone connection.

(3) Radio frequency output power connection.

(4) Audio frequency power amplifier output connector and selector switch.

(5) On-off switch for primary power to transmitter. May be combined with receiver controls such as the receiver on-off switch and volume control.

(6) Upper-lower sideband selector; for single sideband transmitters only.

(7) Selector for choice of carrier level; for single sideband transmitters only. May be combined with sideband selector.

(8) Transmitting frequency selector switch.

(9) Transmit-receive switch.

(10) Meter(s) and selector switch for monitoring transmitter performance.

(11) Pilot lamp or meter to indicate the presence of radio frequency output power or that transmitter control circuits are activated to transmit.

(e) An instruction book for the user shall be furnished with each transmitter sold and one copy (a draft or preliminary copy is acceptable providing a final copy is furnished when completed) shall be forwarded to the Commission with each request for type acceptance or type approval. The book shall contain all information necessary for the proper installation and operation of the transmitter including:

(1) Instructions including all controls, adjustments and switches which may be operated or adjusted without causing violation of technical regulations of this part;

(2) Warnings concerning any adjustment which, according to the rules of this part, may be made only by, or under the immediate supervision of, a person holding a commercial first or second class radio operator license;

(3) Warnings concerning the replacement or substitution of crystals, tubes or other components which could cause violation of the technical regulations of this part and of the type acceptance or type approval requirements of Part 2 of this chapter.

(4) Warning concerning licensing requirements and details concerning the application procedures for licensing.

§ 95.59 *Submission of noncrystal controlled Class C station transmitters for type approval.* Type approval of noncrystal controlled transmitters for use at Class C stations in this service may be requested in accordance with the procedure specified in Part 2 of this chapter.

§ 95.61 *Type approval of receiver-transmitter combinations.* Type approval will not be issued for transmitting equipment for operation under this part when such equipment is enclosed in the same cabinet, is constructed on the same chassis in whole or in part, or is identified with a common type or model number with a radio receiver, unless such receiver has been certificated to the Commission as complying with the requirements of Part 15 of this chapter.

§ 95.63 *Minimum equipment specifications.* Transmitters submitted for type approval in this service shall be capable of meeting the technical specifications contained in this part, and in addition, shall comply with the following:

(a) Any basic instructions concerning the proper adjustment, use, or operation of the equipment that may be necessary shall be attached to the equipment in a suitable manner and in such positions as to be easily read by the operator.

(b) A durable nameplate shall be mounted on each transmitter showing the name of the manufacturer, the type or model designation, and providing suitable space for permanently displaying the transmitter serial number, FCC type approval number, and the class of station for which approved.

(c) The transmitter shall be designed, constructed, and adjusted by the manufacturer to operate on a frequency or frequencies available to the class of station for which type approval is sought. In designing the equipment, every reasonable precaution shall be taken to protect the user from high voltage shock and radio frequency burns. Connections to batteries (if used) shall be made in such a manner as to permit replacement by the user without causing improper operation of the transmitter. Generally accepted modern engineering principles shall be utilized in the generation of radio frequency currents so as to guard against unnecessary interference to other services. In cases of harmful intereference arising from the design, construction, or operation of the equipment, the Commission may require appropriate technical changes in equipment to alleviate interference.

(d) Controls which may effect changes in the carrier frequency of the transmitter shall not be accessible from the exterior of any unit unless such accessibility is specifically approved by the Commission.

§ 95.65 *Test procedure.* Type approval tests to determine whether radio equipment meets the technical specifications contained in this part will be conducted under the following conditions:

(a) Gradual ambient temperature variations from 0° to 125° F.

(b) Relative ambient humidity from 20 to 95 percent. This test will normally consist of subjecting the equipment for at least three consecutive periods of 24 hours each, to a relative ambient humidity of 20, 60, and 95 percent, respectively, at a temperature of approximately 80° F.

(c) Movement of transmitter or objects in the immediate vicinity thereof.

(d) Power supply voltage variations normally to be encountered under actual operating conditions.

(e) Additional tests as may be prescribed, if considered necessary or desirable.

§ 95.67 *Certificate of type approval.* A certificate or notice of type approval, when issued to the manufacturer of equipment intended to be used or operated in the Citizens Radio Service, constitutes a recognition that on the basis of the test made, the particular type of equipment appears to have the capability of functioning in accordance with the technical specifications and regulations contained in this part: *Provided,* That all such additional equipment of the same type is properly constructed, maintained, and operated: *And provided further,* That no change whatsoever is made in the design or construction of such equipment except upon specific approval by the Commission.

§ 95.69 *(Reserved)*
[§ 95.69 deleted eff. 11-23-73; VI (72)-3]

SUBPART D—STATION OPERATING REQUIREMENTS

§ 95.83 *Prohibited uses.* (a) A Citizens radio station shall not be used:

(1) For engaging in radio communications as a hobby or diversion, i.e., operating the radio station as an activity in and of itself.

NOTE: The following are typical, but not all inclusive, examples of the types of communications evidencing a use of Citizens radio as a hobby or diversion which are prohibited under this rule:

"You want to give me your handle and I'll ship you out a card the first thing in the morning;" or "Give me your 10-20 so I can ship you some wallpaper." (Communications to other licensees for the purpose of exchanging so-called "QSL" cards.)

"I'm just checking to see who is on the air."

"Just calling to see if you can hear me. I'm at Main and Broadway."

"Just heard your call sign and thought I'd like to get acquainted;" or "Just passing through and heard your call sign so I thought I'd give you a shout."

"Just sitting here copying the mail and thought I'd give you a call to see how you were doing." (Referring to an intent to communicate based solely on hearing another person engaged in the use of his radio.)

"My 10-20 is Main and Broad Streets. Thought I'd call so I can see how well this new rig is getting out."

"Got a new mike on this rig and thought I'd give you a call to find out how my modulation is."

"Just thought I would give you a shout and let you know I am still around. Thanks for coming back."

"Clear with Venezuela. Just thought I'd let you know I was copying you up here."

"Thought I'd give you a shout and see if you knew where the unmodulated carrier was coming from."

"Just thought I'd give you a call to find out how the skip is coming in over at your location."

"Go ahead breaker. What kind of a rig are you using? Come back with your 10-20."

(2) For any purpose, or in connection with any activity, which is contrary to Federal, State, or local law.

(3) For the transmission of communications containing obscene, indecent, or profane words, language, or meaning.

(4) To carry communications for hire, whether the remuneration or benefit received is direct or indirect.

(5) To communicate with stations authorized or operated under the provisions of other parts of this chapter, with unlicensed stations, or with U.S. Government or foreign stations (other than as provided in Subpart E of this part) except for communications pursuant to § § 95.85(b) and 95.121 and, in the case of Class A stations, for communications with U.S. Government stations in those cases which require co-operation or coordination of activities.

(6) For any communication not directed to specific stations or persons, except for: (i) Emergency and civil defense communications as provided in § § 99.85

(b) and 95.121, respectively, (ii) test transmissions pursuant to § 95.93, and (iii) communications from a mobile unit to other units or stations for the sole purpose of requesting routing directions, assistance to disabled vehicles or vessels, information concerning the availability of food or lodging, or any other assistance necessary to a licensee in transit.

(7) To convey program material for retransmission, live or delay, on a broadcast facility.

Note: A Class A or Class D station may be used in connection with the administrative, engineering, or maintenance activities of a broadcasting station; a Class A or Class C station may be used for control functions by radio which do not involve the transmission of program material; and a Class A or Class D station may be used in the gathering of news items or preparation of programs: *Provided,* That the actual or recorded transmissions of the Citizens radio station are not broadcast at any time in whole or in part.

(8) To interfere maliciously with the communications of another station.

(9) For the direct transmission of any material to the public through public address systems or similar means.

(10) To transmit superfluous communications, i.e., any transmissions which are not necessary to communications which are permissible.

(11) For the transmission of music, whistling, sound effects, or any material for amusement or entertainment purposes, or solely to attract attention.

(12) To transmit the word "MAYDAY" or other international distress signals, except when a ship, aircraft, or other vehicle is threatened by grave and imminent danger and requests immediate assistance.

(13) For transmitting communications to stations of other licensees which relate to the technical performance, capabilities, or testing of any transmitter or other radio equipment, including transmissions concerning the signal strength or frequency stability of a transmitter, except as necessary to establish or maintain the specific communication.

(14) For relaying messages or transmitting communications for a person other than the licensee or members of his immediate family except: (i) Communications transmitted pursuant to § § 95.85 (b), 95.87 (b) (7), and 95.121; (ii) upon specific prior Commission approval, communications between Citizens radio service stations at fixed locations where public telephone service is not provided; and (iii) communications reporting locally observed traffic conditions directed to persons engaged directly or indirectly in furnishing traffic condition information to the motoring public via broadcast facilities.

(15) For advertising or soliciting the sale of any goods or services.

(16) For transmitting messages in other than plain language. Abbreviations, including nationally or internationally recognized operating signals, may be used only if a list of all such abbreviations and their meaning is kept in the station records and made available to any Commission representative on demand.

(b) A Class D station may not be used to communicate with, or attempt to communicate with, any unit of the same or another station over a distance of more than 150 miles.

(c) A licensee of a Citizens radio station who is engaged in the business of selling Citizens radio transmitting equipment shall not allow a customer to operate under his station license. In addition, all communications by the licensee for the purpose of demonstrating such equipment shall consist only of brief messages addressed to other units of the same station.

§ 95.85 *Emergency and assistance to motorist use.* (a) All Citizens radio stations shall give priority to the emergency communications of other stations which involve the immediate safety of life of individuals or the immediate protection of property.

(b) Any station in this service may be utilized during an emergency involving the immediate safety of life of individuals or the immediate protection of property for the transmission of emergency communications. It may also be used to transmit communications necessary to render assistance to a motorist.

(1) When used for transmission of emergency communications certain provisions in this part concerning use of frequencies (§ 95.41(d)); prohibited uses (§ 95.83(a) (5), (6), and (14)); operation by or on behalf of persons other than the licensee (§ 95.87); and duration of transmissions (§ 95.91 (a) and (b)) shall not apply.

(2) When used for transmission of communications necessary to render assistance to a motorist, the provisions of this part concerning directing communications to specific persons or stations (§ 95.83(a) (6)); transmitting messages for other persons (§ 95.83(a) (14)); and duration of transmissions (§ 95.91(b)) shall not apply.

(3) The exemptions granted from certain rule provisions in subparagraphs (1) and (2) of this paragraph may be rescinded by the Commission at its discretion.

(c) If the emergency use under paragraph (b) of this section extends over a period of 12 hours or more, notice shall be sent to the Commission in Washington, D.C., as soon as it is evident that the emergency has or will exceed 12 hours. The notice should include the identity of the stations participating, the nature of the emergency, and the use made of the stations. A single notice covering all participating stations may be submitted.

§ 95.87 *Operation by, or on behalf of, persons other than the licensee.* (a) Transmitters authorized in this service must be under the control of the licensee at all times. A licensee shall not transfer, assign, or dispose of, in any manner, directly or indirectly, the operating authority under his station license, and shall be responsible for the proper operation of all units of the station.

(b) Citizens radio stations may be operated only by the following persons, except as provided in paragraph (c) of this section:

(1) The licensee;

(2) Members of the licensee's immediate family living in the same household;

(3) The partners, if the licensee is a partnership, provided the communications relate to the business of the partnership;

(4) The members, if the licensee is an unincorporated association, provided the communications relate to the business of the association;

(5) Employees of the licensee only while acting within the scope of their employment;

(6) Any person under the control or supervision of the licensee when the station is used solely for the control of remote objects or devices, other than devices used only as a means of attracting attention; and

(7) Other persons, upon specific prior approval of the Commission shown on or attached to the station license, under the following circumstances;

(i) Licensee is a corporation and proposes to provide private radiocommunication facilities for the transmission of messages or signals by or on behalf of its parent corporation, another subsidiary of the parent corporation, or its own subsidiary. Any remuneration or compensation received by the licensee for the use of the radiocommunication facilities shall be governed by a contract entered into by the parties concerned and the total of the compensation shall not exceed the cost of providing the facilities. Records which show the cost of service and its nonprofit or cost-sharing basis shall be maintained by the licensee.

(ii) Licensee proposes the shared or cooperative use of a Class A station with one or more other licensees in this service for the purpose of communicating on a regular basis with units of their respective Class A stations, or with units of other Class A stations if the communications transmitted are otherwise permissible. The use of these private radiocommunication facilities shall be conducted pursuant to a written contract which shall provide that contributions to capital and operating expense shall be made on a nonprofit, cost-sharing basis, the cost to be divided on an equitable basis among all parties to the agreement. Records which show the cost of service and its nonprofit, cost-sharing basis shall be maintained by the licensee. In any case, however, licensee must show a separate and independent need for the particular units proposed to be shared to fulfill his own communications requirements.

(iii) Other cases where there is a need for other persons to operate a unit of licensee's radio station. Requests for authority may be made either at the time of the filing of the application for station license or thereafter by letter. In either case, the licensee must show the nature of the proposed use and that it relates to an activity of the licensee, how he proposes to maintain control over the transmitters at all times, and why it is not appropriate for such other person to obtain a station license in his own name. The authority, if granted, may be specific with respect to the names of the persons who are permitted to operate, or may authorize operation by unnamed persons for specific purposes. This authority may be revoked by the Commission, in its discretion, at any time.

(c) An individual who was formerly a citizens radio station licensee shall not be permitted to operate any citizens radio station of the same class licensed to another person until such time as he again has been issued a valid radio station license of that class, when his license has been:

(1) Revoked by Commission.

(2) Surrendered for cancellation after the institution of revocation proceedings by the Commission.

(3) Surrendered for cancellation after a notice of apparent liability to forfeiture has been served by the Commission.

§ 95.89 *Telephone answering services.*

(a) Notwithstanding the provisions of § 95.87, a licensee may install a transmitting unit of his station on the premises of a telephone answering service. The same unit may not be operated under the authorization of more than one licensee. In all cases, the licensee must enter into a written agreement with the answering service. This agreement must be kept with the licensee's station records and must provide, as a minimum, that:

(1) The licensee will have control over the operation of the radio unit at all times;

(2) The licensee will have full and unrestricted access to the transmitter to enable him to carry out his responsibilities under his license;

(3) Both parties understand that the licensee is fully responsible for the proper operation of the citizens radio station; and

(4) The unit so furnished shall be used only for the transmission of communications to other units belonging to the licensee's station.

(b) A citizens radio station licensed to a telephone answering service shall not be used to relay messages or transmit signals to its customers.

§ 95.91 *Duration of transmissions.* (a) All communications or signals, regardless of their nature, shall be restricted to the minimum practicable transmission time. The radiation of energy shall be limited to transmissions modulated or keyed for actual permissible communications, tests, or control signals. Continuous or uninterrupted transmissions from a single station or between a number of communicating stations is prohibited, except for communications involving the immediate safety of life or property.

(b) Communications between or among Class D stations shall not exceed 5 consecutive minutes. At the conclusion of this 5-minute period, or upon termination of the exchange if less than 5 minutes, the station transmitting and the stations participating in the exchange shall remain silent for a period of at least 5 minutes and monitor the frequency or frequencies involved before any further transmissions are made. However, for the limited purpose of acknowledging receipt of a call, such a station or stations may answer a calling station and request that it stand by for the duration of the silent period. The time limitations contained in this paragraph may not be avoided by changing the operating frequency of the station and shall apply to all the transmissions of an operator who, under other provisions of this part, may operate a unit of more than one citizens radio station.

(c) The transmission of audible tone signals or a sequence of tone signals for the operation of the tone operated squelch or selective calling circuits in accordance with § 95.47 shall not exceed a total of 15 seconds duration. Continuous transmission of a subaudible tone for this purpose is permitted. For the purposes of this section, any tone or combination of tones having no frequency above 150 cycles per second shall be considered subaudible.

(d) The transmission of permissible control signals shall be limited to the minimum practicable time necessary to accomplish the desired control or actuation of remote objects or devices. The continuous radiation of energy for periods exceeding 3 minutes duration for the purpose of transmission of control signals shall be limited to control functions requiring at least one or more changes during each minute of such transmission. However, while it is actually being used to control model aircraft in flight by means of interrupted tone modulation of its carrier, a citizens radio station may transmit a continuous carrier without being simultaneously modulated if the presence of the carrier also performs a control function. An exception to the limitations contained in this paragraph may be authorized upon a satisfactory showing that a continuous control signal is required to perform a control function which is necessary to insure the safety of life or property.

§ 95.93 *Tests and adjustments.* All tests or adjustments of citizens radio transmitting equipment involving an external connection to the radio frequency output circuit shall be made using a nonradiating dummy antenna. However, a brief test signal, either with or without modulation, as appropriate, may be transmitted when it is necessary to adjust a transmitter to an antenna for a new station installation or for an existing installation involving a change of antenna or change of transmitters, or when necessary for the detection. measurement. and suppression of harmonic or other spurious radiation. Test transmissions using a radiating antenna shall not exceed a total of 1 minute during any 5-minute period, shall not interfere with communications already in progress on the operating frequency, and shall be properly identified as required by § 95.95, but may otherwise be unmodulated as appropriate.

§ 95.95 *Station identification.* (a) The call sign of a citizen radio station shall consist of three letters followed by four digits.

(b) Each transmission of the station call sign shall be made in the English language by each unit, shall be complete, and each letter and digit shall be separately and distinctly transmitted. Only standard phonetic alphabets, nationally or internationally recognized, may be used in lieu of pronunciation of letters for

voice transmission of call signs. A unit designator or special identification may be used in addition to the station call sign but not as a substitute therefor.

(c) Except as provided in paragraph (d) of this section, all transmissions from each unit of a citizens radio station shall be identified by the transmission of its assigned call sign at the beginnng and end of each transmission or series of transmissions directed to or exchanged with a unit of the same station or units of other stations. Each required identification shall include not only the call sign of the station unit transmitting, but also the call sign of the station or stations with which the transmitting unit is communicating, or attempting to communicate. In the case of communications between units of the same station (intrastation), after identifying itself by its assigned call sign, the transmitting unit may identify the other units by unit designators. For communications between units of different stations (interstation), the complete sign of all stations involved must be transmitted. If the call sign of the station being called is not known, the name or trade name may be used, but when contact has been made the called station shall thereafter be identified by its call sign. Examples of proper identification procedure are set forth at the end of this paragraph. Where transmissions or exchanges of transmissions of greater length are permitted by this part, the identification shall also be transmitted at least every 15 minutes. Each transmission or exchange of transmissions conducted on different frequencies shall be fully and separately identified in accordance with the foregoing on each frequency used.

Examples of Proper Identification
Intrastation communications:

(1) Calling: "KZZ 0001 base, calling unit 2."
Response: "KZZ 0001 unit 2, to base, over."
Clearing: "KZZ 0001 base, clear with unit 2" and "KZZ 0001 unit 2, clear with base."
(2) Calling: "KZZ 0001 unit 1, calling unit 3."
Response: "KZZ 0001 unit 3, to unit 1, over."
Clearing: "KZZ 0001 unit 1, clear with unit 3, and "KZZ 0001 unit 3, clear with unit 1."
Interstation communications:
Calling: "KZZ 0001 calling KZZ 0002," or "KZZ 0001 calling KZZ 0002 unit 3" (if appropriate).
Response: "KZZ 0002 to KZZ 0001, over."
Clearing: "KZZ 0001 clear with KZZ 0002" and "KZZ 0002 clear with KZZ 0001."

(d) Unless specifically required by the station authorization, the transmissions of a citizens radio station need not be identified when the station (1) is a Class A station which automatically retransmits the information received by radio from another station which is properly identified or (2) is not being used for telephony emission.

(e) In lieu of complying with the requirements of paragraph (c) of this section, Class A base stations, fixed stations, and mobile units when communicating with base stations may identify as follows:

(1) Base stations and fixed stations of a Class A radio system shall transmit their call signs at the end of each transmission or exchange of transmissions, or once each 15-minute period, of a continuous exchange of communications.

(2) A mobile unit of a Class A station communicating with a base station of a Class A radio system on the same frequency shall transmit once during each exchange of transmissions any unit identifier which is on file in the station records of such base station.

(3) A mobile unit of Class A stations communicating with a base station of a Class A radio system on a different frequency shall transmit its call sign at the end of each transmission or exchange of transmissions, or once each 15-minute period of a continuous exchange of communications.

§ 95.97 *Operator license requirements.* (a) No operator license is required for the operation of a citizens radio station except that stations manually transmitting Morse Code shall be operated by the holders of a third or higher class radiotelegraph operator license.

(b) Except as provided in paragraph (c) of this section, all transmitter adjustments or tests while radiating energy during or coincident with the construction, installation, servicing, or maintenance of a radio station in this service, which may affect the proper operaton of such stations, shall be made by or under the immediate supervision and responsibility of a person holding a first- or second-class commercial radio operator license, either radiotelephone or radio telegraph, as may be appropriate for the type of emission employed, and such person shall be responsible for the proper functioning of the station equipment at the conclusion of such adjustments or tests. Further, in any case where a transmitter adjustment which may affect the proper operation of the transmitter has been made while not radiating energy by a person not the

holder of the required commercial radio operator license or not under the supervision of such licensed operator, other than the factory assembling or repair of equipment, the transmitter shall be checked for compliance with the technical requirements of the rules by a commercial radio operator of the proper grade before it is placed on the air.

(c) Except as provided in § 95.53 and in paragraph (d) of this section, no commercial radio operator license is required to be held by the person performing transmitter adjustments or tests during or coincident with the construction, installation, servicing, or maintenance of Class C transmitters, or Class D transmitters used at stations authorized prior to May 24, 1974: *Provided,* That there is compliance with all of the following conditions:

(1) The transmitting equipment shall be crystal-controlled with a crystal capable of maintaning the station frequency within the prescribed tolerance;

(2) The transmitting equipment either shall have been factory assembled or shall have been provided in kit form by a manufacturer who provided all components together with full and detailed instructions for their assembly by nonfactory personnel;

(3) The frequency determining elements of the transmitter, including the crystal(s) and all other components of the crystal oscillator circuit, shall have been preassembled by the manufacturer, pretuned to a specific available frequency, and sealed by the manufacturer so that replacement of any component or any adjustment which might cause off-frequency operation cannot be made without breaking such seal and thereby voiding the certification of the manufacturer required by this paragraph;

(4) The transmitting equipment shall have been so designed that none of the transmitter adjustments or tests normally performed during or coincident with the installation, servicing, or maintenance of the station, or during the normal rendition of the service of the station, or during the final assembly of kits or partially preassembled units, may reasonably be expected to result in off-frequency operation, excessive input power, overmodulation, or excessive harmonics or other spurious emissions; and

(5) The manufacturer of the transmitting equipment or of the kit from which the transmitting equipment is assembled shall have certified in writing to the purchaser of the equipment (and to the Commission upon request) that the equipment has been designed, manufactured, and furnished in accordance with the specifications contained in the foregoing subparagraphs of this paragraph. The manufacturer's certification concerning design and construction features of Class C or Class D station transmitting equipment, as required if the provisions of this paragraph are invoked, may be specific as to a particular unit of transmitting equipment or general as to a group or model of such equipment, and may be in any form adequate to assure the purchaser of the equipment or the Commission that the conditions described in this paragraph have been fulfilled.

(d) Any tests and adjustments necessary to correct any deviation of a transmitter of any Class or station in this service from the technical requirements of the rules in this part shall be made by, or under the immediate supervision of, a person holding a first- or second-class commercial operator license, either radiotelephone or radiotelegraph, as may be appropriate for the type of emission employed.

§ 95.101 *Posting station license and transmitter identification cards or plates.* (a) The current authorization, or a clearly legible photocopy thereof, for each station (including units of a Class C or Class D station) operated at a fixed location shall be posted at a conspicuous place at the principal fixed location from which such station is controlled, and a photocopy of such authorization shall also be posted at all other fixed locations from which the station is controlled. If a photocopy of the authorization is posted at the principal control point, the location of the original shall be stated on that photocopy. In addition, an executed Transmitter Identification Card (FCC Form 452-C) or a plate of metal or other durable substance, legibly indicating the call sign and the licensee's name and address, shall be affixed, readily visible for inspection, to each transmitter operated at a fixed location when such transmitter is not in view of, or is not readily accessible to, the operator of at least one of the locations at which the station authorization or a photocopy thereof is required to be posted.

(b) The current authorization for each station operated as a mobile station shall be retained as a permanent part of the station records, but need not be posted. In addition, an executed Transmitter Identification Card (FCC Form 452-C) or a plate of metal or other durable substance, legibly indicating the call sign and the licensee's name and address, shall be affixed, readily visible for inspection. to each of such transmitters: *Provided,* That, if the transmitter is not in view of the location from which it is controlled, or is not readily accessible for inspection, then such card or plate shall be affixed to the control equipment at the transmitter operating position or posted adjacent thereto.

§ 95.103 *Inspection of stations and station records.* All stations and records of stations in the Citizens Radio Service shall be made available for inspection upon the request of an authorized representative of the Commission made to the licensee or to his representative (see § 1.6 of this chapter). Unless otherwise stated in this part, all required station records shall be maintained for a period of at least 1 year.

§ 95.105 *Current copy of rules required.* Each licensee in this service shall maintain as a part of his station records a current copy of Part 95, Citizens Radio Service, of this chapter.

§ 95.107 *Inspection and maintenance of tower marking and lighting, and associated control equipment.* The licensee of any radio station which has an antenna structure required to be painted and illuminated pursuant to the provisions of section 303(q) of the Communications Act of 1934, as amended, and Part 17 of this chapter, shall perform the inspection and maintain the tower marking and lighting, and associated control equipment, in accordance with the requirements set forth in Part 17 of this chapter.

§ 95.111 *Recording of tower light inspections.* When a station in this service has an antenna structure which is required to be illuminated, appropriate entries shall be made in the station records in conformity with the requirements set forth in Part 17 of this chapter.

§ 95.113 *Answers to notices of violations.* (a) Any licensee who appears to have violated any provision of the Communications Act or any provision of this chapter shall be served with a written notice calling the facts to his attention and requesting a statement concerning the matter. FCC Form 793 may be used for this purpose.

(b) Within 10 days from receipt of notice or such other period as may be specified, the licensee shall send a written answer, in duplicate, direct to the office of the Commission originating the notice. If an answer cannot be sent nor an acknowledgement made within such period by reason of illness or other unavoidable circumstances, acknowledgement and answer shall be made at the earliest practicable date with a satisfactory explanation of the delay.

(c) The answer to each notice shall be complete in itself and shall not be abbreviated by reference to other communications or answers to other notices. In every instance the answer shall contain a statement of the action taken to correct the condition or omission complained of and to preclude its recurrence. If the notice relates to violations that may be due to the physical or electrical characteristics of transmitting apparatus, the licensee must comply with the provisions of § 95.53, and the answer to the notice shall state fully what steps, if any, have been taken to prevent future violations, and, if any new apparatus is to be installed, the date such apparatus was ordered, the name of the manufacturer, and the promised date of delivery. If the installation of such apparatus requires a construction permit, the file number of the application shall be given, or if a file number has not been assigned by the Commission, such identification shall be given as will permit ready identification of the application. If the notice of violation relates to lack of attention to or improper operation of the transmitter, the name and license number of the operator in charge, if any, shall also be given.

§ 95.115 *False signals.* No person shall transmit false or deceptive communications by radio or identify the station he is operating by means of a call sign which has not been assigned to that station.

§ 95.117 *Station location.* (a) The specific location of each Class A base station and each Class A fixed station and the specific area of operation of each Class A mobile station shall be indicated in the application for license. An authorization may be granted for the operation of a Class A base station or fixed station in this service at unspecified temporary fixed locations within a specified general area of operation. However, when any unit or units of a base station or fixed station authorized to be operated at temporary locations actually remains or is intended to remain at the same location for a period of over a year, application for separate authorization specifying the fixed location shall be made as soon as possible but not later than 30 days after the expiration of the 1-year period.

(b) A Class A mobile station authorized in this service may be used or operated anywhere in the United States subject to the provisions of paragraph (d) of this section; *Provided,* That when the area of operation is changed for a period exceeding 7 days, the following procedure shall be observed:

(1) When the change of area of operation occurs inside the same Radio District, the Engineer in Charge of the Radio District involved and the Commission's office, Washington, D.C., 20554, shall be notified.

(2) When the station is moved from one Radio District to another, the Engineers in Charge of the two Radio Districts involved and the Commission's office, Washington, D.C., 20554, shall be notified.

(c) A Class C or Class D mobile station may be used or operated anywhere in the United States subject to the provisions of paragraph (d) of this section.

(d) A mobile station authorized in this service may be used or operated on any vessel, aircraft, or vehicle of the United States; *Provided,* That when such vessel, aircraft, or vehicle is outside the territorial limits of the United States, the station, its operation, and its operator shall be subject to the governing provisions of any treaty concerning telecommunications to which the United States is a party, and when within the territorial limits of any foreign country, the station shall be subject also to such laws and regulations of that country as may be applicable.

§ 95.119 *Control points, dispatch points, and remote control.* (a) A control point is an operating position which is under the control and supervision of the licensee, at which a person immediately responsible for the proper operation of the transmitter is stationed, and at which adequate means are available to aurally monitor all transmissions and to render the transmitter inoperative. Each Class A base or fixed station shall be provided with a control point, the location of which will be specified in the license. The location of the control point must be the same as the transmitting equipment unless the application includes a request for a different location. Exception to the requirement for a control point may be made by the Commission upon specific request and justification therefor in the case of certain unattended Class A stations employing special emissions pursuant to § 95.47(e). Authority for such exception must be shown on the license.

(b) A dispatch point is any position from which messages may be transmitted under the supervision of the person at a control point who is responsible for the proper operation of the transmitter. No authorization is required to install dispatch points.

(c) Remote control of a citizens radio station means the control of the transmitting equipment of that station from any place other than the location of the transmitting equipment, except that direct mechanical control or direct electrical control by wired connections of transmitting equipment from some other point on the same premises, craft, or vehicle shall not be considered remote control. A class A base or fixed station may be authorized to be used or operated by remote control from another fixed location or from mobile units: *Provided,* That adequate means are available to enable the person using or operating the station to render the transmitting equipment inoperative from each remote control position should improper operation occur.

(d) Operation of any Class C or Class D station by remote control is prohibited.

§ 95.121 *Civil defense communications.* A licensee of a station authorized under this part may use the licensed radio facilities for the transmission of messages relating to civil defense activities in connection with official tests or drills conducted by, or actual emergencies proclaimed by, the civil defense agency having jurisdiction over the area in which the station is located: *Provided,* That:

(a) The operation of the radio station shall be on a voluntary basis.

(b) [Reserved]

(c) Such communications are conducted under the direction of civil defense authorities.

(d) As soon as possible after the beginning of such use, the licensee shall send notice to the Commission in Washington, D. C., and to the Engineer in Charge of the Radio District in which the station is located, stating the nature of the communications being transmitted and the duration of the special use of the station. In addition, the Engineer in Charge shall be notified as soon as possible of any change in the nature of or termination of such use.

(c) In the event such use is to be a series of pre-planned tests or drills of the same or similar nature which are scheduled in advance for specific times or at certain intervals of time, the licensee may send a single notice to the Commission in Washington, D. C., and to the Engineer in Charge of the Radio District in which the station is located, stating the nature of the communications to be transmitted, the duration of each such test, and the times scheduled for such use. Notice shall likewise be given in the event of any change in the nature of or termination of any such series of tests.

(f) The Commission may, at any time, order the discontinuance of such special use of the authorized facilities.

SUBPART E—OPERATION OF CITIZENS RADIO STATIONS IN THE UNITED STATES BY CANADIANS

§ 95.131 *Basis, purpose and scope.* (a) The rules in this subpart are based on, and are applicable solely to the agreement (TIAS #6931) between the United States and Canada, effective July 24, 1970, which permits Canadian stations in the General Radio Service to be operated in the United States.

(b) The purpose of this subpart is to implement the agreement (TIAS #6931) between the United States and Canada by prescribing rules under which a Canadian licensee in the General Radio Service may operate his station in the United States.

§ 95.133 *Permit required.* Each Canadian licensee in the General Radio Service desiring to operate his radio station in the United States, under the provisions of the agreement (TIAS #6931), must obtain a permit for such operation from the Federal Communications Commission. A permit for such operation shall be issued only to a person holding a valid license in the General Radio Service issued by the appropriate Canadian governmental authority.

§ 95.135 *Application for permit.* (a) Application for a permit shall be made on FCC Form 410-B. Form 410-B may be obtained from the Commission's Washington, D.C., office or from any of the Commission's field offices. A separate application form shall be filed for each station or transmitter desired to be operated in the United States.

(b) The application form shall be completed in full in English and signed by the applicant. The application must be filed by mail or in person with the Federal Communications Commission, Gettysburg, Pa. 17325, U.S.A. To allow sufficient time for processing, the application should be filed at least 60 days before the date on which the applicant desires to commence operation.

(c) The Commission, at its discretion, may require the Canadian licensee to give evidence of his knowledge of the Commission's applicable rules and regulations. Also the Commission may require the applicant to furnish any additional information it deems necessary.

§ 95.137 *Issuance of permit.* (a) The Commission may issue a permit under such conditions, restrictions and terms as it deems appropriate.

(b) Normally, a permit will be issued to expire 1 year after issuance but in no event after the expiration of the license issued to the Canadian licensee by his government.

(c) If a change in any of the terms of a permit is desired, an application for modification of the permit is required. If operation beyond the expiration date of a permit is desired an application for renewal of the permit is required. Application for modification or for renewal of a permit shall be filed on FCC Form 410-B.

(d) The Commission, in its discretion may deny any application for a permit under this subpart. If an application is denied, the applicant will be notified by letter. The applicant may, within 30 days of the mailing of such letter, request the Commission to reconsider its action.

§ 95.139 *Modification or cancellation of permit.* At any time the Commission may, in its discretion, modify or cancel any permit issued under this subpart. In this event the permittee will be notified of the Commission's action by letter mailed to his mailing address in the United States and the permittee shall comply immediately. A permittee may, within 30 days of the mailing of such letter, request the Commission to reconsider its action. The filing of a request for reconsideration shall not stay the effectiveness of that action, but the Commission may stay its action on its own motion.

§ 95.141 *Possession of permit.* The current permit issued by the Commission, or a photocopy thereof, must be in the possession of the operator or attached to the transmitter. The license issued to the Canadian licensee by his government must also be in his possession while he is in the United States.

§ 95.143 *Knowledge of rules required.* Each Canadian permittee, operating under this subpart, shall have read and understood this Part 95, Citizens Radio Service.

§ 95.145 *Operating Conditions.* (a) The Canadian licensee may not under any circumstances begin operation until he has received a permit issued by the Commission.

(b) Operation of station by a Canadian licensee under a permit issued by the Commission must comply with all of the following:

(1) The provisions of this subpart and of Subparts A through D of this part.

(2) Any further conditions specified on the permit issued by the Commission.

§ 95.147 *Station identification.* The Canadian licensee authorized to operate his radio station in the United States under the provisions of this subpart shall identify his station by the call sign issued by the appropriate authority of the government of Canada followed by the station's geographical location in the United States as nearly as possible by city and state.

CHAPTER 4

HOW TO BUY A CB RADIO

Now that you think you know it all and are ready to go out and plunk down your hard-earned money for a transceiver, where do you start? How do you choose? Do you plan to install it yourself? Or do you need the help of an expert installer?

Many stores selling C.B. radio equipment today provide absolutely no service, no installation and no customer information. These are mainly the big discount stores and mass merchandisers who'll sell you a set of bath towels in the department right next to the C.B. radio display. If you need help, stay away from the mass merchants and make up your mind that you'll have to spend the few extra bucks at a hobby type or radio specialists store. After all, these are the guys who've been selling C.B. equipment since 1958, and they know something about the goods.

Price Points. Every product line has what are called price points; these are the prices at which the equipment is meant to sell retail. For example, a C.B. AM transceiver typically has three price points: $119.95–$129.95; $149.95–$169.95; and $189.95–$199.95 (see Fig. 4-1). The first is called "low-end;" the middle is called "step-up," and the last is "tip-of-the-line" or "deluxe model." Generally, when a store runs a sale with a special, low price on C.B.s, it's the low-end model, and they'll knock off from $10 to $20. It's not unusual for the set to be out of stock by the time you get there, since C.B. radios are often in short supply for certain makes and models. Besides, each store goes on the premise that once the customer is there, he can be easily moved to a step-up model, and with good reason; the step-up unit has features that you'll find desirable.

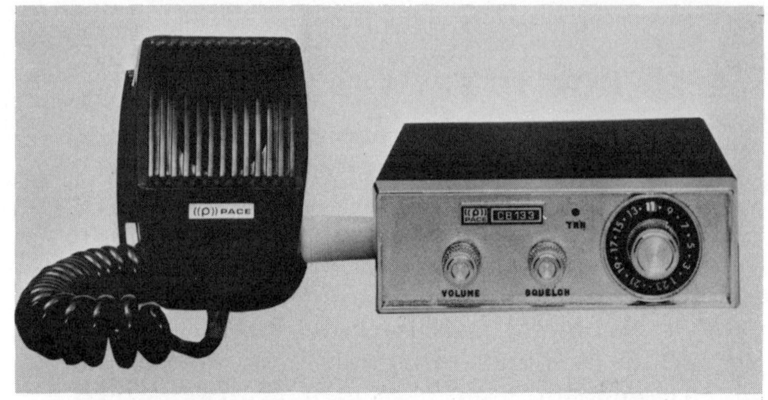

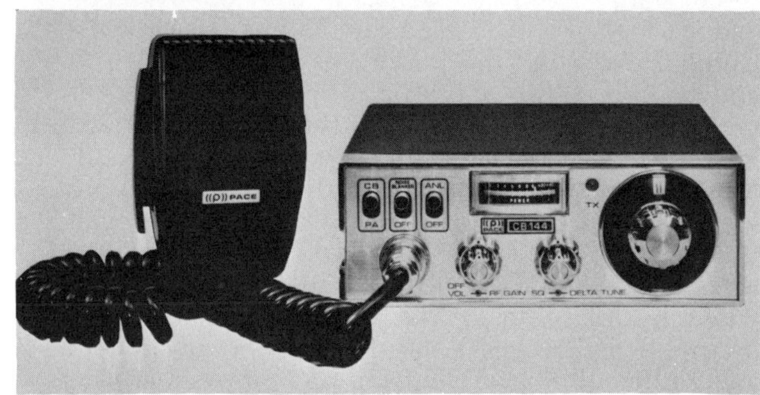

4-1 Typical lineup showing three price points—low-end (A); step-up (B) and high-end (C). These units are typical of the wide range of equipment available from suppliers like Pace/Pathcom.

COURTESY PATHCOM, INC.

Low-End. The low-end transceiver has all 23 channels, an on/off volume control and a squelch control. It may or may not have a public address feature. If it does, it's probably switched on by a position on the channel selector, between channels 22 and 23. It also may or may not have an S/RF meter. The back of the set will have accessory jacks for the PA speaker and an extension speaker.

The step-up model will have an S/RF meter, Delta tune control, and probably a switch for the automatic noise limiter and/or the noise blanker (see Fig. 2). The deluxe model will have both ANL and NB switches, probably a separate switch for the PA, there will be transmit and receive indicator lamps, and possibly an RF gain control. The meter will be larger and easier to read, and there may be such special features as back-of-the-set relay contacts for turning off your car radio or stereo when you're transmitting, additional accessory jacks and other bells and whistles.

4-2 This single sideband base station by Royce (Model 1-640) is loaded with extra-cost features and controls—all desirable for a permanent, one-site installation.

All three sets in the model line have the same basic electrical characteristics: they have 23-channel operation and they deliver close to a four-watt output to the antenna. Some features on the more expensive sets may be hidden. These can include special circuitry that provides optimum modulation, and noise blankers that have no external switching. Some radios have ultra-sensitive receivers with crystal lattice filters and some sets provide RF gain switching for local/distant stations.

Is It All Necessary? Now that you've made up your mind just how many bells and whistles you're willing to pay for, consider next where you're going to put the mobile rig. There are just so many places you can install a CB radio in your car. Some cars don't give you much choice. If the set has to be mounted some distance from the driver, will you be able to see and change the channel selector easily without running the car into a telephone pole? Can you reach, identify and operate *all* controls by feel? Does the microphone plug into the side or the front? Side-mounted mikes can create problems when the set is installed.

Some recent sets have large LED digital numerals (see Fig. 3) which are easy to read from any location in the family buggy. Sets with this feature are becoming more numerous.

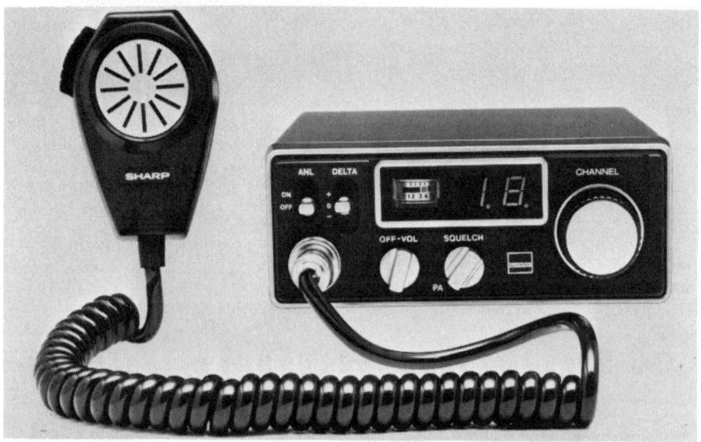

4-3 Large, self-illuminated LED display shows channel numbers on this C.B. transceiver by Sharp.

Another arrangement to consider seriously is an in-dash combination unit—one that combines C.B. radio with AM/FM stereo and possibly a tape player (see Fig. 4). While such a radio is terribly complicated and can create servicing problems, manufacturers are getting better at design, and in many cases now provide the C.B. operating controls on the microphone, to cut down dashboard control clutter.

4-4 Great-grandaddy of all in-dash combination units, this model, the 852 C.B. from J.I.L., combines 23-channel C.B., AM/FM stereo and eight-track tape cartridge player all in a single unit.

In-dash combo units typically sell for $300 to $400, and have most of the same features that are found on underdash C.B.-only radios. Operating convenience is an important factor. Make sure you can work all those tiny knobs and switches by feel before deciding to go this route.

FCC Restrictions. All C.B. radios are designed to meet certain basic legal restrictions imposed by the Federal Communications Commission. The maximum output power to the antenna is four watts. This is true whether the set is a bargain special for $99.95 or a super-deluxe job with a digital clock and weather scan. The same restriction is made on the base station you buy for your home: four watts is the top.

To get more out of the radio, consider the antenna installation. For your first try, you'll probably go to a simple trunk-lip mount antenna for the car since it's relatively inexpensive ($25–$35) and easy to install (see Fig. 5). But pairs of phased antennas such as those used on trucks and recreational vehicles add a little extra charge to your output signal (see Fig. 6). In any event, get a package price for your system when you buy—transceiver, antenna and installation. Some stores will throw in the antenna or the installation free, or for a nominal charge—that is, if they're charging you full list price (or near to it) for the radio. This is the specialist store's way of giving you a discount.

4-5 Typical trunk-lip mount antenna from Hy-Gain attaches to trunk lip quickly, requires no holes for installation.

4-6 Twin phased antennas mounted on truck mirrors help the trucker's C.B. reach in this Pace installation.

Single Sideband. One way to get additional mileage out of a C.B. radio is with a single sideband. This mode of transmission is much more efficient than straight AM, and the FCC rules permit a maximum of 12 watts PEP (peak envelope power) in SSB mode. Also, by going sideband, it's possible, theoretically at least, to triple the number of available channels, since each of the 23 channels has three operational modes—AM, upper sideband, and lower sideband (see Fig. 7).

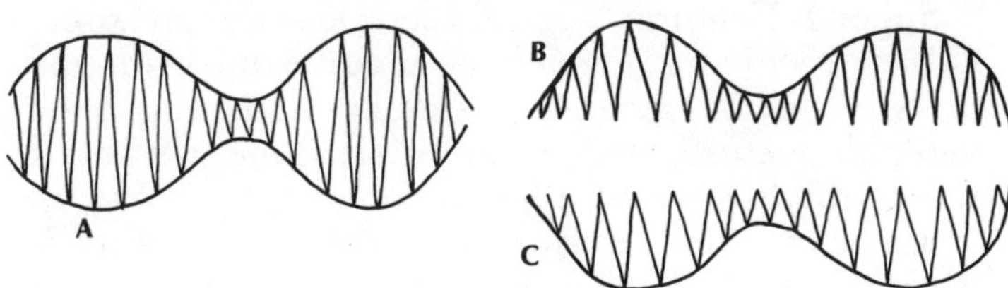

4-7 (A) Full AM modulated carrier with both sidebands present. (B) Upper sideband only and (C) lower sideband.

Single sideband transceivers cost more than the highest price AM radio—typically about $300 to $350. They are physically larger than the straight AM transceivers and have a jillion controls on them (see Fig. 8). You'll see such strange names as "clarifier," which is a continuous fine tuning control, instead of Delta tune. There is often a tone control, and most SSB sets have an RF gain control. If you decide to go sideband, remember that the number of SSB sets out there on the road is relatively small, and you may end up operating in straight AM at four watts much of the time.

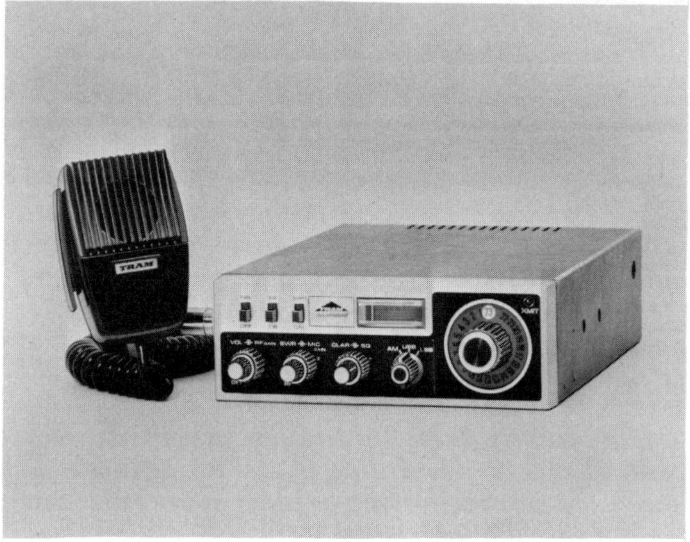

4-8 Compact single-sideband mobile radio by Tram/Diamond.

Special Features. An added feature on some C.B. transceivers is weather—receive only—on one or two special receive channels. This can be a valuable feature, since most entertainment-type car radios don't have weather receive. This tunes in to the U.S. Weather Bureau station, which broadcasts constant weather reports, usually from major airports. Weather broadcasts are available now in

52

many parts of the country. The government is opening up new facilities all the time. Hopefully at some point in the near future, most of the continental United States will have continuous weather reports.

Some C.B. radios are equipped with *two-channel scan*, a valuable feature if you want to use one channel while monitoring another one for example, emergency channel nine. Channel nine is usually quiet since it's against the law to use it for anything except emergencies and travelers assistance. The problem is that not too many people on the road listen in to this channel, because there's so little action on it. By having the scan feature on your transceiver, it's possible to keep Channel nine on at all times while talking and listening on Channel 19 or any other frequency you choose.

One problem with the scan radios: the second channel usually has a "priority" feature—it will switch over to nine even when you're listening to an important message on some other channel, making you miss the message. Of course, it won't switch over while you're transmitting. Still, it can be a valuable feature—certainly a lifesaver in some situations.

Don't Overbuy. One mistake beginning C.B. owners may make is to overbuy. Because a particular set advertised at a loss-leader price is out of stock, don't let yourself be talked up to the "step-up" model that has features you aren't prepared to use. If you've already made up your mind that you want a no-frills model to start, stick with that concept, even if the store can't supply the advertised special.

You may also be seduced by the overabundance of accessory items that are available. More about power microphones, external test meters, external speakers, C.B. headphones and other such, later. Use the radio first. Get on the air and get a feel for

C.B. first. You'll know soon enough which accessories you really need.

Specmanship Simplified. If you're lucky enough to get hold of a specification sheet or a detailed advertisement for the transceiver you plan to buy, you can review some of the C.B. radio's key features before you buy. But what do all those numbers and features mean? This is where specmanship comes in—the ability to interpret all those numbers and often overinflated claims. It also means developing the ability to separate real features from advertising copywriters' flights of fancy.

A typical spec sheet may say something like this (author's comment in parentheses):

- Compact size (most C.B.s are).
- Full 23 channels complete with crystals for each channel (almost all C.B. sets today provide all 23 channels, and with the synthesizer circuits they use, there is no such thing as separate crystals for each channel).
- Beautiful walnut grain cabinet (so what?)
- Large, easy-to-read meter (they're never big enough)
- Squelch control (every C.B. set has one).
- Automatic noise limiter (most C.B. rigs have this circuit; step-up models usually have defeat switch)
- Fine-tuning control (or Delta tune—helpful, in some cases for off-frequency incoming signals.)
- PA switch (almost all C.B. sets have PA setting; many have it as a position on the channel selector dial between channels 22 and 23.)
- RF gain control (or local/distance switch which cuts down the receiver's sensitivity for close-by stations to decrease overload distortion)
- Volume control with on/off switch (so show us a C.B. set that doesn't have this)
- Convenient microphone jack (you've got to plug in the mike *somewhere*)

- Detachable electronic microphone included (Would you buy a set without one?)
- Built-in speaker (ho-hum)
- Dependable channel selector (dependable until the contacts get dirty)
- Solid state (nobody uses tubes in mobile rigs anyway, so they're *all* solid-state.)
- Floating chassis for positive/negative ground (featured on most mobile rigs today to eliminate problems associated with mounting in foreign cars)
- Maximum 4-watt power (important, since some manufacturers still mention only the old, now-outmoded FCC specification of 5-watt *input*)
- Modulation and receive indicators (useful as a check that you're actually on the air)
- Built-in automatic modulation control (not all sets have an effective circuit for this; it generally gives 90 percent modulation for safe, but clear transmission)
- FCC license form included (if it's not, you can pick up a copy from the dealer)
- External speaker jack (most all sets have this, for plugging in a larger, cleaner sounding external speaker)
- PA jack (to plug in that PA speaker)
- Antenna connector (just try to work without an antenna!)
- Directly wired power cord (for the juice that makes the set work)

All the preceding specs come under the heading of "a little information and a lot of advertising puff to fill up the spec sheet." Now here are some *real* specs (see Fig. 9):

- Audio power output 2.5 watts (enough to drive any external speaker and a PA speaker)
- Sensitivity 0.5 μV (also written 0.5 uV—the

Greek *mu* means "micro." 0.5 μV means "half of a microvolt"—a very good figure for C.B. AM radios. Generally, anything *less* than 1.0 μV is good for this kind of radio; the smaller the number, the better.)

- Transmitter power 4 watts (under old rules, 5-watt input to the final was the maximum allowed, but due to transmitter inefficiency, the actual output to the antenna was often as little as 2 watts. Now sets are defined by the actual transmitter power output.)
- Modulation control set for 90 percent (this means that you'll get plenty of talk power, but may later want to invest in a power microphone to boost this level to 90 or 100 percent. Anything over 100 percent causes distortion.)
- Current drain, receive at max. audio, 1.3 A; standby 0.5 A (this tells you how much of a drain the set will be on your car's battery. This is important if you do much C.B. operating from the roadside with the engine off.)
- Fine tuning range ±1.5 kHz (the fine tune or Delta tune control can raise or lower the received frequency by 1.5 kHz.)

Before you get discouraged, you don't have to know the meanings of all those terms and numbers; the data is given here merely as a guide to help you get started in the right direction. It is to demonstrate that some manufacturers play games with their specifications, and the idea is not to get taken in by spectacular claims. This is especially true of transmitter power, since all C.B. radios, whether mobile or base station, are limited by law to 4 watts output.

Supplies may be short, which means that you may not be able to buy the particular C.B. radio you've set your heart on. Be equipped with at least a mental list of alternates or their equivalents when you go to the store. There are many fine C.B. radios on the market today, and they're bargains, even at full list price.

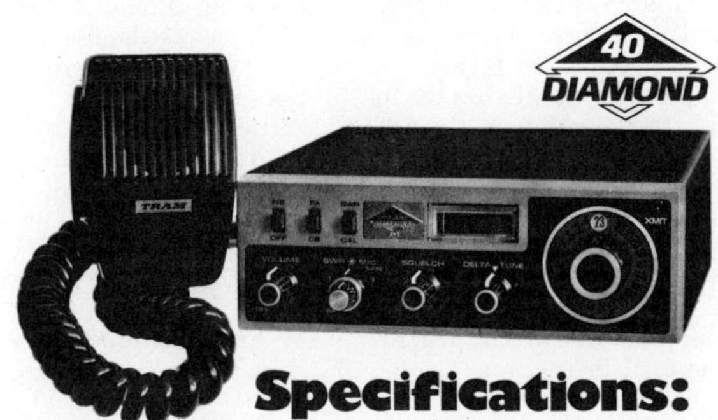

RECEIVER SPECIFICATIONS

Sensitivity:	.3uV provides more than 1 watt audio. Signal modulated — 30% @ 1kHz sine wave.
Signal to Noise Ratio:	—.3uV for 10db S+N/N, signal mod. 30% at 1kHz sine wave.
Selectivity:	6db @ 4kHz, 60db @ 20 kHz.
AGC:	Change in audio output less than 12db from 10uV to .5 volts.
Squelch:	Adjustable — Threshold less than .5uV. Tight more than 100uV.
Audio Freq. Response:	500 to 2000Hz.
Distortion:	Less than 10% at 5.0 watts output.
Image Rejection:	More than 50db.
IF Rejection:	More than 80db at 455kHz.
Adjacent Channel Rejection:	More than 60db at .3uV.
Cross Modulation:	More than 55db.
IF Frequency:	10.00–10.04mHz, 455kHz
Clarifier Range:	± 1.5kHz.
Noise Blanker:	RF parallel gate type (hybrid IC).

TRANSMITTER SPECIFICATIONS

AM Power:	5 watts.
Modulation Capability:	100%.
Harmonic Suppression and Spurious Emissions:	Better than FCC requirement.
Frequency Response:	500Hz to 3kHz ± 5db.

GENERAL SPECIFICATIONS

Frequency Tolerance:	0.005%.
Frequency Stability:	0.001%.
Operating Temperature Range:	−30°C to +50°C.
Microphone:	Dynamic with push-to-talk switch and coiled cord.
Supply Voltage:	13.8V DC (Positive or negative ground).
Current Drain:	Receive — 1.3A @ Max. audio output. 0.5A standby. (no signal)
Meter:	Illuminated, indicates receiving signal strength, relative power output and SWR.
Size:	2 3/8'' (H) x 6 5/8'' (W) x 8 1/16'' (D).
Weight:	6.0 pounds.

A DESIGN PRODUCT OF DIAMOND MICROWAVE CORPORATION. EACH AND EVERY SET IS 100% QUALITY CONTROLLED AND POWER CHECKED AT DIAMOND'S U.S.A. FACTORY.

4-9 Typical spec sheet from Tram/Diamond. The other side contains layman's information, but no real specs.

"TENNIS SHOES" Shines. . . .

Between LA and Vegas, the Nevada desert stretches in all directions, lonely and forbidding, with nothing but an occasional piece of sagebrush to keep you company. To the stranded motorist, it can cause acute anxiety symptoms.

I had such symptoms one hot, blazing day as I stepped out of my station wagon and squinted about, shading my eyes against the glare of the sun. I had stopped by the side of the road to relax and enjoy a cup of coffee from my thermos. Wanting to leave, my attempts to start the car ended in failure. Plenty of gas, and I was as familiar with mechanics as any educated layman. A repeated check of the engine had failed to produce the reason for its stubborn refusal to turn over.

I hated the inconvenience, but wasn't worried too much since I knew that I could always rely on my CB radio. I switched to Channel 9, "React Monitor, React Monitor, this is KWF 4121 with a 10-33. . . . React Monitor, do you copy me. . . ." Crackles and the rushing noise of the radio. Otherwise: Silence. "React Monitor, I need help." I repeated my message several times. No reply. Only a faint level of skip met my words.

Abandoning Channel 9, I repeated my cry for help through each channel. Same disheartening results. Finally, to my vast relief, I heard a conversation in progress on Channel 20. "Break! Break!" I called breathlessly.

Almost immediately, back came a welcome response. "Breaker, this is Big Mama. What can I do for you?"

Gratefully I stated my problem and gave my location as shown on the nearest highway marker. The voice took down the information in a very professional manner, reassured me that my fate was in good hands and signed off. After another moment I heard "Tennis Shoes, Tennis Shoes, do you read Big Mama?"

"Big Mama, this is Tennis Shoes. What's up?"

"We got one right up your alley," explained the voice from Big Mama. "A car stuck on the highway, just over the hill from you. Thought you might like to help out."

"Sure thing," returned Tennis Shoes, "I got a stray calf out yonder anyhow. Might's well kill two birds."

"OK, Tennis Shoes, and to Vegas mobile, we'll call AAA. But if she gets you started first, let us know."

She? I wondered, but replied with thanks. Feeling a sudden warm glow of friendship for the Nevada wilderness, I settled back with a magazine to pass the time. Presently, a labored chug-chug broke the silence. I looked up. A dusty, ancient Model T Ford was pulling up beside me. My eyes widened in amazement as out popped a sprightly little old lady, seventy years old, if she was a day, clad in a knee-length skirt and bright-red sweater, complete with tennis shoes and bobby socks. Her grey hair was pulled back in a long braid from her sun-browned wrinkled face. Someone's gotta be kidding, I thought.

"You the car in trouble?"

I nodded, speechless.

"I'll just take a look at that there engine, then." The grey head disappeared under the raised hood. I heard a thumping and a ping, saw one sun-browned hand extract a bobby pin from the grey braid, heard another jiggle or two, and a scrape. "Turn'er over now, Sonny," I heard her call.

Without much hope, I pushed in the starter. To my utter amazement, the engine caught and roared to life.

The old lady slammed the hood shut in satisfaction. "That'll fix'er," she chortled triumphantly. "Better keep 'er runnin' for now, though, Sonny," she instructed, "til you get to Winnemucca. Then pull into the first Shell station on your right and tell Joe you got a Tennis Shoes Special. He'll know what to do." I was simply astounded as she waved I stammered thanks aside. "Ya better git goin' now."

Turning, she hopped into her car and nursed its wheezing motor to life. As I pulled away, I heard through my CB, "Big Mama, this is Tennis Shoes. Call off the AAA boys and chalk up another bobby pin victory. I'll keep an eye on him 'til he gits out 'o sight, then go look for that dang calf. 73 and I'm out."

I have lived in Pasadena for years. Never thought much of "little old ladies with tennis shoes." They were always holding up traffic. "Clunk Clunking" in their old cars. You were never sure what their next move or turn would be. But from now on don't ever say anything mean about "little old ladies in tennis shoes" when I'm around.

I am also keeping a bobby pin handy. Not that I know what to do with it. But some "little old lady in tennis shoes" might.

COURTESY PACE/PATHCOM, INC.
(PACE COMMUNICATIONS DIVISION)

COURTESY PATHCOM, INC.

CHAPTER 5

INSTALLING THE EQUIPMENT

Base Station. Your base station operates on a local antenna. Sometimes your base station unit comes with a small vertical whip antenna that plugs into the back of the unit and all you have to do to install is to plug it into an electrical wall outlet. All at once you are on the air. However, this type of installation is not the best since you are not going to get the maximum efficiency that can be obtained with a proper outdoor antenna.

What is best, is to install your antenna, mounted on stand-off insulators at the peak of a roof. These insulators are porcelain devices which keep the antenna from touching the wood. Now, you run your coaxial feeder cable from the antenna to the transceiver. If you run it outside the house, down along one of the window lines and then bring it in through a near-by window, you are going to have everything blowing in the wind and slapping against the house making noise. This is not going to do the wire or the connector any good.

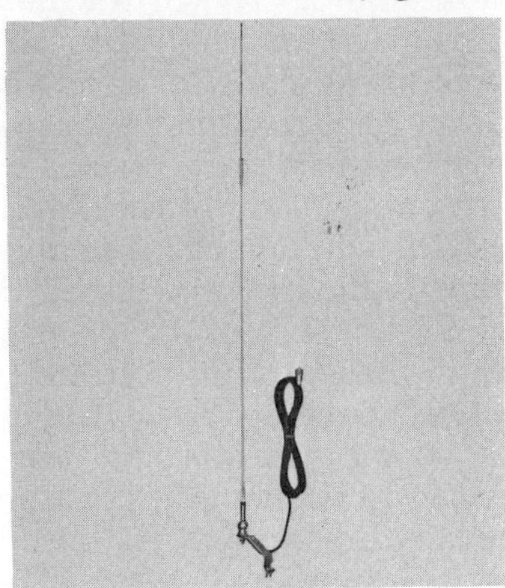

5-1 Trunk-groove mounted 48″ center-loaded antenna.

In your local radio supply store you will find nails about four or five inches long with loops at one end. These kind of nails are called "stand-off". Run your wire through them. They hold the wire rigidly in place against the side of the house because you bang the nails directly into the house. You can also get these nails for masonry or brick.

Bring the wire down to the floor on which you are going to put your unit. Now the question arises: how are you going to bring the antenna wire from outside to the inside?

One way is to take a piece of wood exactly the width of your window sash. Drill a hole in it, run the wire through the hole, pull the window down on the wood, and then seal the balance of the opening where the wire comes through with some putty or caulking compound. Afterward, you can paint the whole thing to match the rest of the house.

Another, especially good way of doing it, if you have a shingled roof, is to purchase a device called a feed-through. This consists of a big plastic base which is very light in weight. The hole is recessed under a rain shield. You drill a hole through the roof near the antenna and slip the plastic plate under one of the shingles. Then tack it down with shingle tacks. In your attic or crawl space, run the wire through, caulk it, and you will have the rest of the wire running inside of the house. Always try to run the wire parallel and close the the AC power lines in your wall. By doing so, it makes the job much easier.

Now, in a base station, you are going to work with a critical length of coaxial feedline from the transmitter to the antenna and one of the things you are going to want to do is to prune the feed-line. The feedline carries the radio energy from transceiver to antenna. If the feedline is of a certain length it is liable to absorb the energy instead of transmitting it to the antenna.

5-2   COURTESY GC ELECTRONICS

There is a device called a *balun,* or "bazooka". It is an *impedance matching transformer* that you can put in and not have to worry about "pruning" the antenna. Or, you can get a *field strength meter.* Turn on the transmitter and watch how high up the needle jumps. Then begin clipping the feedline shorter and shorter, taking notice of whether the needle goes up or down. When it begins to go up, start clipping shorter and shorter pieces so that you will not take too much off at one time. When it starts to drop down just the least little bit, stop. That means most of the energy is going up to the antenna.

Now that we are on the subject of base stations and their installation, have you considered where you are going to put your unit? You'll want to put it where it is most convenient to the purpose of owning it. For example, if you want to use it mostly for business you probably would put it in the den. Or, you might locate it to conform with where you are most of the time.

5-2A Quick release C.B. antenna bracket is anti-theft and easy to install.

It's always a good idea to mount your unit a little
bit high up if there are children in the house. In
this way they won't be tempted to play with it.
Remember that you are responsible, even if it is
your children who break the law unknowingly. To
the Federal Communications Commission it is still
your fault.

5-3 SWR/WATT Meter relays antenna system's performance.

Thanks to *push-to-talk microphones* and a selector switch it is possible to get a separate microphone hanger. This is a device to which the microphone is clipped. Instead of having to reach up to hang it on the side of the cabinet, you can actually put it anywhere that is most convenient. Don't forget the transmitter is part of this, too, because you are operating simply by pushing to talk.

5-4 Microphone with gain controlled amplifier uses standard 9-volt battery and has push-to-talk function.

COURTESY MURA CORPORATION

Base stations are designed with what looks like a carrying handle. It certainly can be used as such if you are in need. But if you fold it underneath the unit it serves as an easel support so that the unit tilts up toward you if you are using it as a desk or table. However, you can remove this handle. You

will find screw holes in it. Mount it under a shelf
and then put the unit back in so that the unit rests
underneath it. You will find this to be very convenient in the long run.

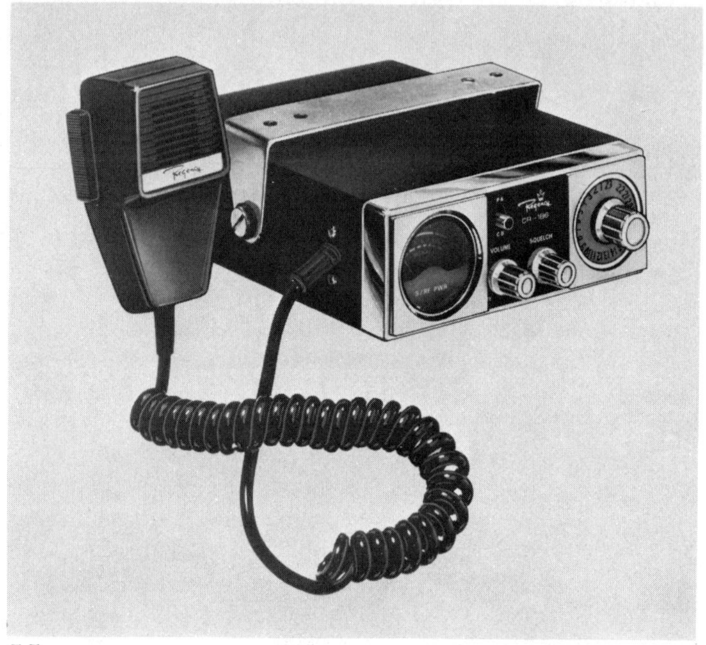

5-5

Mobile Installation. In the mobile installation,
you would most probably want to mount your rig
underneath the dashboard. This is done with a
carrying handle device which is attached to a couple of holes in the dashboard. For making a permanent installation, loosen the two screws that
hold the chassis and panel to the case at the back
of your transceiver. Remove the insides, leaving
the empty case.

At the top of the case there are a couple of holes
and two larger ones at the bottom of the case, per-
pendicular and opposite to these. The holes at the

bottom permit you to stick a screwdriver through the bottom and tighten the screws into the dashboard.

Now, there might be one or two extra holes in the bottom of your dashboard, but let us assume that there aren't any. Take a pencil and stick it through the top and bottom holes. Hold the unit in place under your dashboard, where you are going to mount it, and make two little circles, using the top holes as guides. With a number twenty-four drill, drill two small holes. Use number six self-tapping screws to permanently attach the case under the dashboard.

You might also want to put an anchor strap at the reel of the case to support the back end of the transceiver to prevent it from wobbling. Once you've got the case properly mounted beneath the dash, slip the guts back in, tighten the two screws at the back. Your installation is almost finished.

The next thing you will have to do is your electrical connecting. If you have mounted the case properly to the metal dashboard of the car, then the case is at ground potential. It's grounded. You will not have to connect a ground wire. There will be a wire coming out of the back with a fuse in it. The other end of that wire has to connect to the twelve-volt battery of the car.

Now, this does not have to connect directly to the battery but can be joined with any twelve-volt source. The best thing to do is find your fuse block, where there is usually an empty connector to which you could attach this wire and thus pick up the 12 volts.

In the event you do not want to do this, open up the hood of the car and look at the fire wall, which is the panel between the passenger compartment and the engine compartment. Find a suitable hole, punch it through with a screwdriver, and run the wire through that. Then peek at your car's battery. You will find that one wire goes to the chassis. That is the ground wire. The other wire goes either to the alternator or to the regulator. There you will be able to loosen a screw and attach the twelve volt wire and with very little trouble, too. The only thing you have left to do is to put your antenna in place.

Location of Loudspeaker. There are certain things that you should consider when you are installing your equipment. For instance, where is the loudspeaker on your transceiver? Is it on the side of the rig? If it is, then make certain you keep that side open. In other words, do not move that side of the transceiver right up against the glove compartment or it will muffle the sound.

Is the loudspeaker on top of your unit? Well, then you must make sure that you have a suspended installation using a hang-over device rather than mounting the unit flush up against the dashboard. If the loudspeaker is on top and the unit is underneath your dashboard, then the sound

5-6

68

is going to go right under the dashboard. You will never hear anything this way and you will probably jump to the conclusion that everybody is giving off a bad signal.

So it is best to have the speaker right up front. Most of the sets are designed that way today, anyway. Make sure you mount where you will have easy access while you are driving your car. You might want to flip from channel to channel and you must make certain that the microphone is within easy reach, and that you will be able to lift it right out of its holder. Also make sure that all the controls on the transceiver are within easy reach.

Another and very important thing to consider is making sure you will mount your unit in such a way that it will not interfere with the functioning of the automobile.

Now, there are people who feel they do not want to mar their car in the slightest way. They do not even want to use a chain bumper mount. For these people, two alternatives are available.

One is a suction cup, about four to five inches in diameter. It looks like a plumber's helper. You wet this with glycerin and place it on top of the roof of the car. This gives a ground-plane effect antenna because the body of the car acts like a reflector. Wire is run through an open window and in this way the car is completely unharmed and it is also easy to take off.

Another type of mount is an antenna that clips right on to the rain gutter with two small screws. This also does no damage to the car. This same method can be used on still another type of mount in which you open up the trunk of the car and clip the antenna to the body of the car underneath the open trunk. Closing the trunk locks it in place.

Now, you might ask, "Will I ever have to demount the unit?" The answer is yes. And keep this in mind when you are mounting it. There is

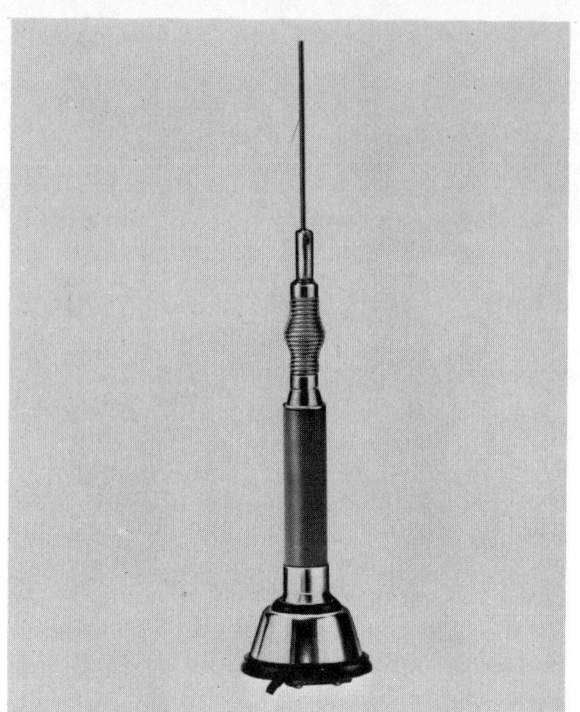

5-7 Base loaded trunk-lid or roof-top antenna for mobile C.B. transceivers.

COURTESY MURA CORPORATION

5-8 Antenna rain gutter mount.

COURTESY NEW-TRONICS CORP.

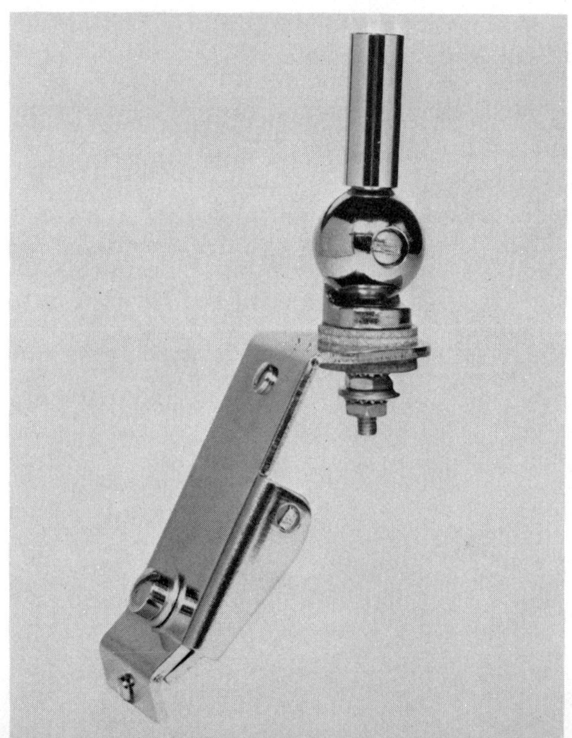

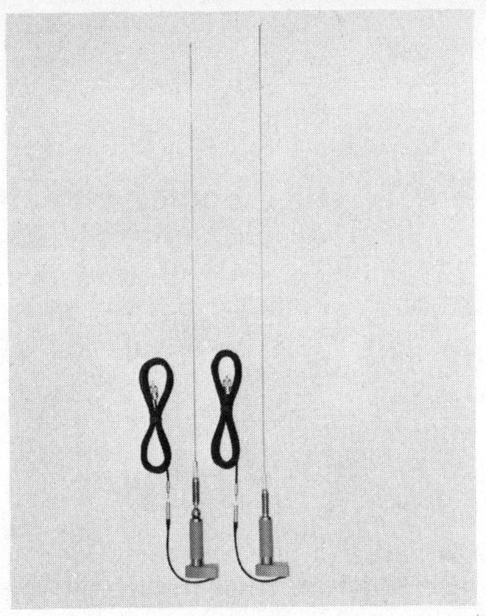

5-9 Trunk lip mount 48″ mobile antennas.
COURTESY NEW-TRONICS CORP.

Servicing the Mobile Unit. If your unit ever needs servicing, it will certainly have to come out. So you will have to mount with the thought that should the occasion ever arise, it could be removed without causing a great deal of trouble.

In the case of an underdash mount, all you would have to do is unscrew the unit, disconnect the wires, and just pull it out. You do not need the case anyway, when you are having it serviced. It is only a decoration when you get down to it.

And also take under consideration that if you bought a unit with crystals for four channels and it had enough room for twenty-three channels, that you might in the future think of adding additional ones. In that case, you would have to remove the set in order to put in those extra crystals.

Other Types of Installation. What about the safety of the unit? Suppose you were parked on a city street and the set is sitting in your car in plain sight? It could be awfully tempting to steal. If someone knew how to mount it, it would certainly be no task for him to demount it.

Well, there is still another type of mount in which the transceiver is mounted in the glove compartment. So if yours will fit, and if you have the space for it, go to it. You can install your own rig without any skill and without specialized tools. Manufacturers provide full installation instructions with clear diagrams.

Actually, installing Citizens Band equipment is a pretty pat thing to do. It consists of deciding whether or not it is to be installed as a base station or as a mobile station, and finding the place where you want it put in, mechanically mounting the transceiver and the antenna, and then connecting the antenna to the transceiver and the transceiver to the voltage source.

There is something called the two-way installation. Suppose you have a base station but you also want to use it in your car. This requires that when you mount it in the car it should be mounted in such a way as to be easily removed to bring back into the house.

Some units are designed with a double case. They are equipped with a case that is large and can be mounted in the car and a smaller case that is mounted on the unit. In this way you can carry the unit back into the house, using it the way it is, leaving the outer case in the car. To use it in the car you just lip it in. This can be very convenient if you are driving in an area in which you would rather not leave the unit in the car. Why, you can carry it right along with you.

Installing Antenna in a Car. And so, giving due consideration to how and where you want to install your equipment, you are now ready for the installation. How you are going to install your antenna depends on the kind you are going to use. Since the vertical whip is the most common, we will begin with that.

The vertical whip antenna usually mounts at the

back end of your car because you want to get it as far away from the car's big metal body as possible. There are quite a few mounting methods you can use.

There is a type of chain mount that attaches to the rear bumper of your car. It locks in place with an ordinary wrench. Make certain it is as tight as you can get it. When it is properly installed and

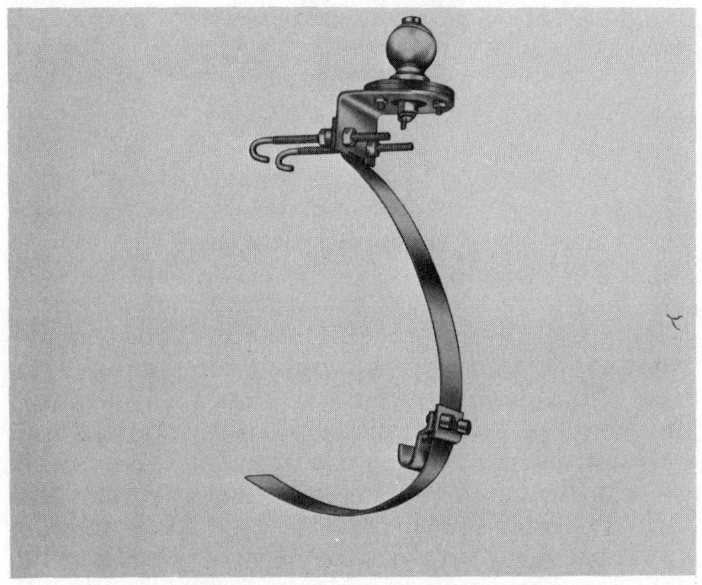

5-10 CB mobile antenna bumper mount.

clamped on to the bumper, there is a threaded portion to which you attach this big black plastic ball. This ball is cut on a diagonal and has a sawtoothed surface on both of its sides.

Notice, that by rotating the upper part of the ball you can change the angle of the car's antenna. Make it absolutely straight because it will compensate if you mounted the bumper mount either too far forward or too far back. But you can still straighten out the antenna by rotating the black plastic ball.

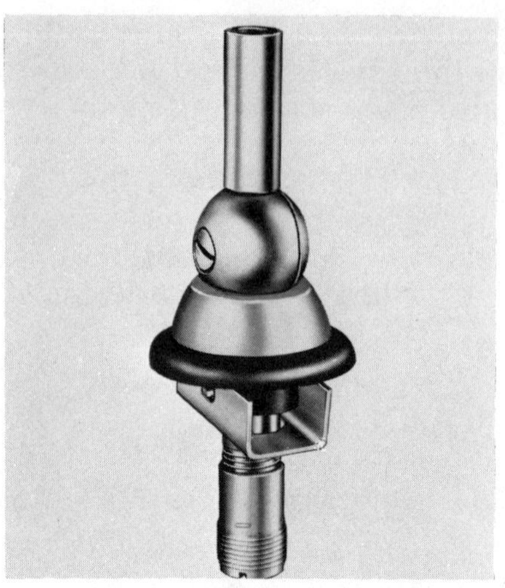

5-11 CB mobile antenna cowlmount.

All right! The ball is mounted. Now take the antenna whip and screw it right on to the top of the ball. This done, you have to find a way to bring the wire into the trunk compartment of the car. You will need to punch a hole in the body of the car near the trunk. There is an easy way to do this. Drill a quarter-inch hole. You can take a punch or a nail and bang a little dent into the surface of the vehicle so you will be able to drill right through more easily. You can use a tapered reamer or a quarter-inch drill to enlarge the hole.

Next, get a three-piece device called a *chassis punch*. Insert it into the drilled hole and tighten it. It will punch out a neat and clean one-half-inch hole. Insert a rubber *grommet* into the opening. This will fit snugly around the wire going to the inside of the trunk.

If you intend to use a body mount, perhaps on the side of your fender, you would work in pretty much the same way. You have to make a punch mark where you want the hole to be placed, but

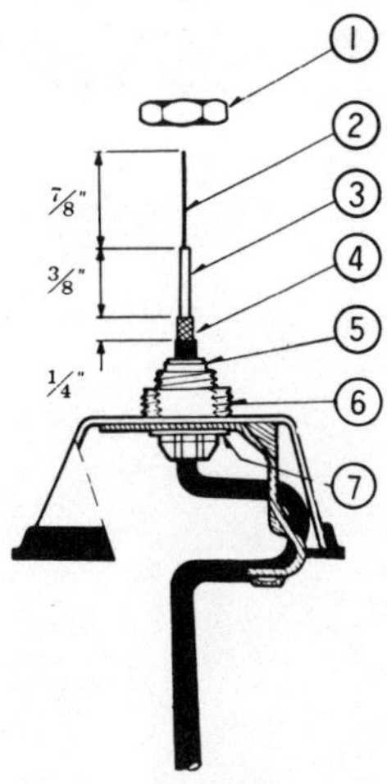

1	NUT
2	COPPER CENTER CONDUCTOR
3	INNER INSULATION
4	FLEXIBLE BRAIDED COPPER SHIELD
5	SLEEVE
6	MOUNTING SCREW
7	MOUNTING LOCK WASHER

this time drilling out a three-eighth-inch opening. This will require a larger chassis punch, one and one-eighth inches in diameter, depending upon the antenna.

Proceed to punch out the hole, disassemble the antenna mount, place it in the opening, and then bolt the other end of it at the back from inside the car so that the antenna mount is held solidly on the side of the automobile. You will find another one of these ball contraptions to enable you to set the vertical whip so it remains vertical and not sticking out at all angles.

When you have it set, tighten so it stays firmly in place. Now screw the antenna to it. Look at it! Check it out with your eye! Does it look straight to you? You know, with this type of mount you do not need another hole and that is because your antenna wire has already been fed through the hole to the body of your car.

Okay! Over the door of your car, near the front, there is a gutter that keeps rain from running down. You need a gutter clip here. A gutter clip is a metal or plastic device that attaches with two screws to this gutter. A hooker is then shaped so

when you bend your antenna forward it will latch
onto this hook, thus keeping it bent over so you
can drive your car in and out of the garage without
the antenna getting smashed up.

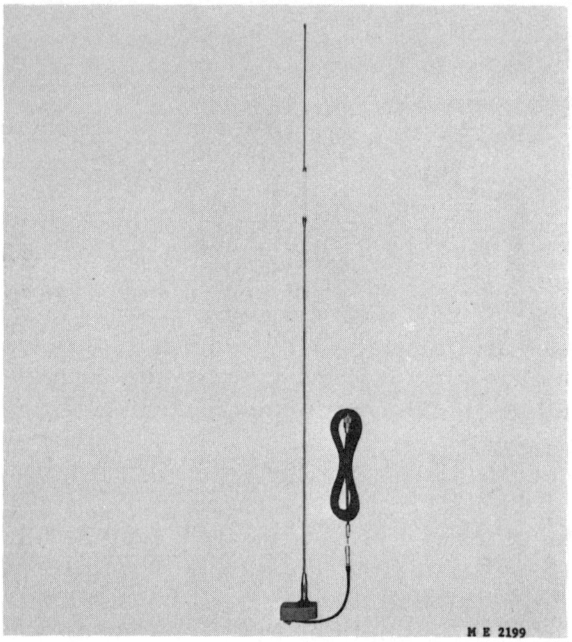

5-13 48″ super flex stainless steel trunk lip mount antenna
COURTESY NEW-TRONICS CORP.

There is still another type of antenna called the
loaded-coil antenna. The C.B. antenna must be of
a given critical length. The length is determined
by the frequency on which it operates. When you
are receiving, the longer and higher the antenna is,
the better off you are. When you are transmitting,
the length of the antenna is most critical. For C.B.,
it is usually about eight feet long.
If you feel unhappy about having such a long antenna, you can wind five feet of it around a coil.
You can buy a coil having five feet of heavy gauge
wire which has been wound and sealed in a plastic

76

tube. Take one foot of antenna, screw this coil to the top of it and then take two more feet of antenna and screw this to the top of the coil.

In reality, it is one and a half feet on either side, making the total antenna length three feet. However, the effective length is still eight feet. This is called a center-loaded antenna.

There are other forms of this kind of antenna. One of them is the base-loaded antenna in which the coil goes directly into the car, and to the mount. Then when the antenna is placed on top of that, you have your eight feet.

Although there are many variations of this kind of antenna, the effect is usually the same. The antenna is effectively shortened without decreasing its performance electronically.

Making Connections. Okay! Now that the transceiver and antenna are mounted, the next thing to do is to connect them. Of course, you will want to do this properly, neatly, and as easily as possible. Let us begin on the driver's side of the car.

Remove the trim along the side of the door—the part that keeps the carpeting in place. There you will see some Phillips-head screws. Remove them and lift that piece of trim away.

Connect the antenna wire to the transceiver. Underneath the dashboard, you will find cables and wires which are held in place with cable clamps. Instead of undoing the cable clamps, attach the antenna wire to any of these wires and then leave it. You can attach it in place with some tape or waxed thread to keep it out of the way.

Lie on your back on the front seat of your car with your head underneath the dashboard to see what you are doing. A good strong lantern could be very helpful here. Lift up the carpeting from where you took the trim away and then lay the cable in underneath the carpeting along the side of the car.

Open your back door and take the trim off the

bottom of it. You will notice a piece of plastic trim that goes from the doorpost between the two doors to over the carpeting on the floor. Loosen this and remove it. Now run the wire all the way back, just as far as you possibly can, to the rear portion of the back door under the carpet.

At this particular point, you ought to be able to replace the trim from the front door, locking the carpeting back in its place. Next, lift out the back seat of your car and remove it from the vehicle all together.

Run the wire underneath the seat area and open up the trunk. With the trunk open, door closed, and you on your back on the floor, you should be able to see daylight. You are actually looking through a hole in the compartment between the trunk and the passenger section of the car.

Suppose you do see such a hole. All right! Then run your wire up to and through the opening which would place the wire right in the trunk.

But suppose you do not see such a hole. Very well! Then drill one. Do not worry about how this is going to look because nobody is ever going to have the pleasure of seeing it. So, once you get the wire into the trunk, you can replace the trim on the rear door and also replace the back seat.

Now, go around to the back of the car and pull the wire into the trunk, taking up all the slack. Then make your connection from that wire to the antenna which has now been fed through the body of the car. At this point, the installation is complete. Turn on your rig and try it out.

Base Station Chimney Mount Antenna. Mr. Richard Weiss of Mountain Communications, Orange, New Jersey recommends a base station chimney mount because the chimney is the highest and sturdiest place to install an antenna.

According to Mr. Weiss, the first thing to do is select a suitable mounting location. Next, unroll the straps of the chimney mount and stretch them around the chimney. In the back of the straps are two securing brackets with holes in them. Insert the bolt through the holes. Attach nut. Hand tighten first to set and line up. Then, evenly tighten the nuts securely. The mount is on the chimney.

The next step is to mount the antenna to the chimney mount. Hoist the antenna up and hold the mast against the mounting brackets. Slip the U-Bolt over each strap of the chimney mount. Fasten U-Bolts using the nuts supplied. Now tighten them securely. Your antenna is all mounted!

MOBILE INSTALLATION

Mount the "U" bracket supplied with the FS-3023 under the dash panel using the
bracket itself as a location template. Drill holes for screws of the self-tapping
type. If possible, use bolt, nut and lockwasher. A hole is supplied on the rear
panel of the FS-3023 for an additional mounting strap. The use of a wing bolt o
the rear will facilitate easy removal when servicing is necessary. Do not crowd
the FS-3023. Leave room for ventilation.

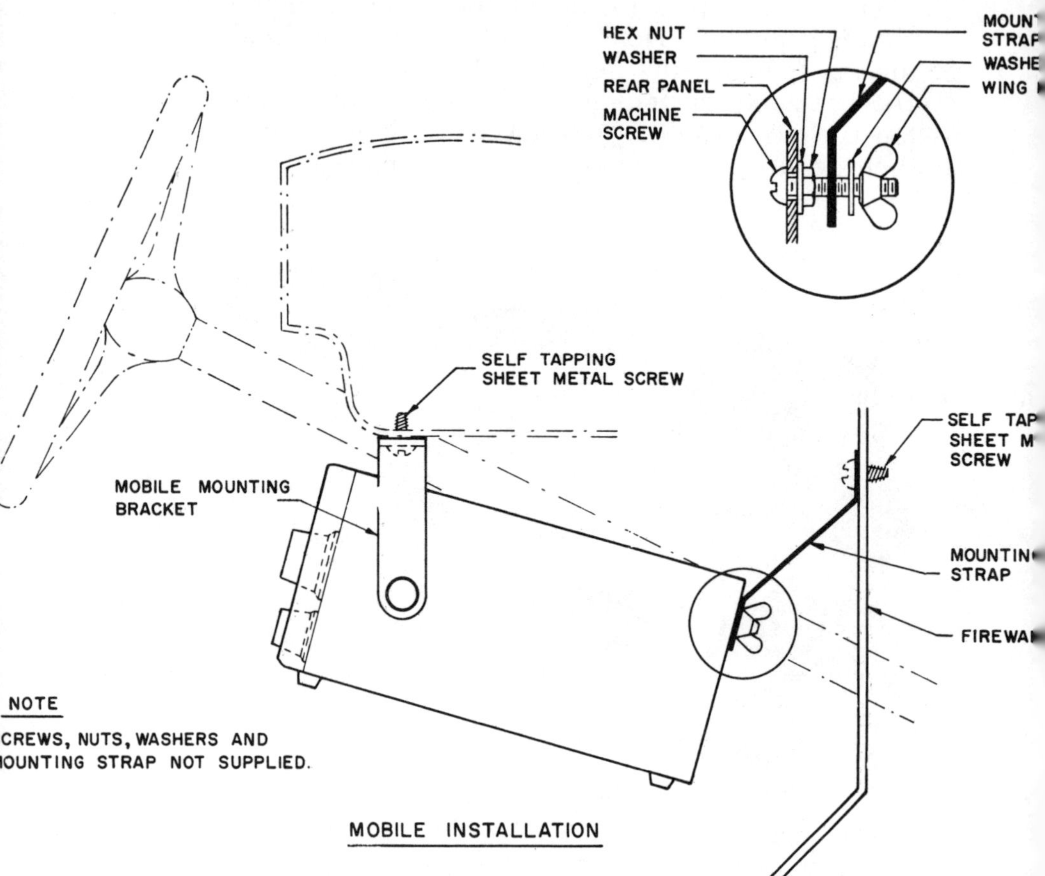

DC POWER - NEGATIVE GROUND ONLY!

The FS-3023 power plug polarity must be carefully noted.   The ground of the
FS-3023 is negative. Use No. 8 to 12 stranded wire for power leads. Be sure
that voltage drops in the power leads are held to a minimum. Red designates
the positive lead of the pre-wired DC power socket and cable. No. 12 wire is
sufficient for 6 foot leads and No. 10 wire for 15 foot leads.

NOISE SUPPRESSION

The FS-3023 Gated Noise Limiter is very effective in the reduction of noise, but this alone is not effective against all the different noises in a mobile installation.

a. Bonding: The use of 1" copper braid to interconnect parts of the automobile that can radiate noise such as:

   (1) Hood to firewall
   (2) Rear bumper to body and chassis
   (3) Rear light fixtures to body
   (4) Tailpipe to body
   (5) Either side of muffler to body and chassis
   (6) Chassis to body in several places
   (7) Ignition coil body to firewall

When braid connections are made, be sure to clean the metal "bright" and coat with grease before tightening the connection. This will prevent contact corrosion which is the chief cause of noise.

A very good method for locating noisy fixtures is to put the FS-3023 into operation and connect a 25' length of coaxial cable to the antenna connector on the rear of the FS-3023. The other end of the coaxial cable should have the center conductor showing for about 1/2". This bare end of the coax will serve as a "noise probe." With the receiver volume turned up and noise limiter "off", touch the "noise probe" to all parts of the automobile (except the electrical system). A large increase in noise will indicate a noisy section. This section should then be bonded and rechecked. Continue this process until a substantial reduction of noise is achieved. REMEMBER! Ungrounded metal parts can radiate noise!

b. Ignition Radiation Suppression requires the use of resistor spark plugs, feed-thru capacitors and distributor suppressors. Of prime importance is a properly adjusted ignition system. The following steps will serve as a guide:

   (1) Spark Plugs: Install resistor spark plugs or Belden IRS cable.
   (2) Distributor Cap: Install suppressor resistor or IRS cable between distributor cap and ignition coil.
   (3) Generator: Install a 0.5 mfd coaxial capacitor (Sprague #48018 or equivalent) at the "A" terminal of the generator.
   (4) Alternators: Require no attention except when the diodes become defective.
   (5) Ignition Coil Primary: Install a 0.1 mfd coaxial capacitor (Sprague #48P9 or equivalent) in the lead from ignition switch to coil. Keep capacitor close to coil terminal. Brighten the metal around the coil mounting bracket to engine block, apply grease and retighten mounting screws.

    (6)(a) <u>Regulator Field Terminal:</u> Connect a 39 ohm resistor in series
with a 0. 01 mfd ceramic capacitor between the Field terminal and
ground.

      (b) <u>Armature Terminal</u>: Insert 0. 2 mfd coaxial capacitor (Sprague
#48P18 or equivalent).

      (c) <u>Battery Terminal</u>: Repeat (b).

  (7) <u>Gauges</u>: Install 0. 5 mfd, 200 volt capacitors from terminals to ground.

  (8) <u>Wheels and Tires</u>: Inject special graphite powder into the tires available
at automotive parts suppliers.

<u>CA</u>UTION: Do not connect any capacitor alone from the Field Terminal of the
generator to ground. Read (6)(a) carefully.

<u>MOBILE ANTENNA</u>

The mobile antenna represents an electrical quarter wave-length at the operating
frequency or physically represents 109". Shorter equivalents are the "loaded"
type of antenna. This type of antenna can be bottom, center or top-loaded and is
usually 5 feet in length. Another loaded type is the spiral-wound antenna.

The best type without a doubt is the 109" whip antenna. This is usually mounted
on the rear bumper or low on the rear of the body. The shorter antenna is a
compromise between efficiency and size, but in most cases can be relied upon to
perform very well. It is very important that the coaxial antenna lead be between
14 to 18 feet to provide a low standing-wave ratio. RF-58A/U cable is recommended as a low-loss transmission line.

There are antenna mounts available to suit any installation. The most practical
types are the swivel ball mount which is available for body mounting, and the
coiled spring type which is used as a bumper mount. In certain instances, both
the swivel ball and coil spring are used.

<u>MARINE ANTENNA</u>

The antenna system for a boat requires a ground plane antenna.   The common
name for such an antenna is "coaxial ground plane antenna. " The boat ground
system which usually consists of a ground plate is not efficient at 27 MHz and,
therefore, should not be relied upon for use with a simple whip antenna. This
does not hold true if the boat is of all metal construction.

The same bonding and ignition suppression techniques must be applied to a boat
as to an automobile. In many cases a boat requires more work because the engine compartment is of wood, whereas an automobile has a hood and firewall to
shield the engine. A boat's wooden engine compartment requires copper mesh
shielding that is adequately connected to a bonded electrical system.

Bonding a boat's ground requires that all metal fittings that come in contact with
the electrical system or water be continuously interconnected by 2" wide copper
"flashing" strips. The engine shaft will require a "wiper" resting on the shaft
and connected to the bonded ground system. This "wiper" is usually a piece of
spring steel resting on a cleaned portion of the shaft.

BASE ANTENNA

transmission line (antenna lead) of a Base installation may be lengthy even
ugh the antenna is only 20' off a roof.   In such an installation RG-8U cable
avier than RG-58A/U) should be used to minimize the transmission line losses.

se antenna falls into two categories:

(a) Ground Plane types
(b) Beam antenna types

und plane antennas are omni-directional.   Beam antennas are uni-directional.
he FS-3023 is used in a restricted direction, a Beam will have a greater ad-
tage since the antenna response will be concentrated in only one direction. The
enna should have an impedance of 50 ohms regardless of the type chosen.

COURTESY SONAR RADIO CORPORATION

CHAPTER 6

HOW TO SELECT AN ANTENNA

There are a number of things to consider when you are buying an antenna for your rig. Although your C.B. unit comes with an instruction book that recommends one antenna or another, you ought to be aware of the fact that in many cases, the C.B. manufacturer is paid a certain amount of money by the antenna manufacturer to recommend his antenna.

And, you should be aware that the antenna length for C.B. is the same for all C.B. units. In other words, you do not not need a certain type of antenna for a certain C.B. rig. It's the same thing as when your television antenna can cover the television frequency spectrum.

But even knowing this, there are still other considerations when selecting an antenna, such as what it is made of. For example, some antennas are made of steel. You hear the word steel and you immediately come to the conclusion that it must be the greatest. But it does have a drawback. Steel tends to rust.

Suppose you have white siding on your house and you put up a steel antenna on it. Well, the house can become stained from the rust marks. And what if you put it on a car that has a good finish on it? The same thing will happen. You'll end up with rust marks on your car.

Some people buy steel antennas just the same because you can get one at a lower price.

Some antennas are made of fiberglass. These are fitted with a steel core so that the fiberglass protects the steel. And, there are also aluminium antennas on the market, with aluminium rods.

As we discussed in the previous chapter, there are beam antennas you can use on your home if you want a directional signal. Then there is the

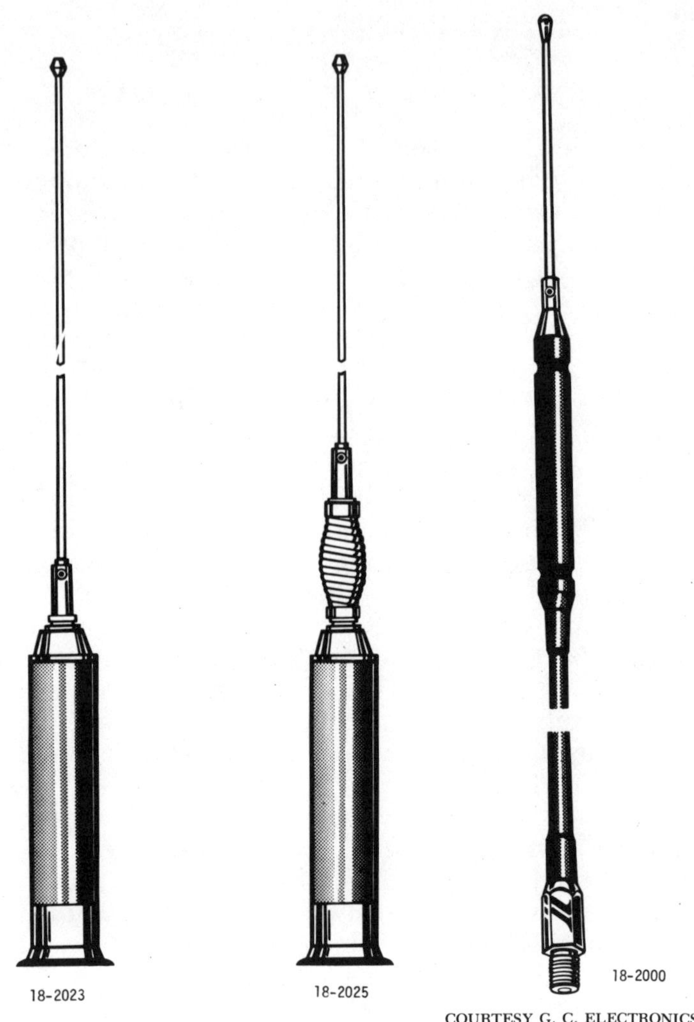

vertical whip. You should keep in mind that this vertical antenna radiates a donut-shaped pattern with no signal at the top and no signal at the bottom. All the signal is coming out of the sides in a three-hundred and sixty degree circle.

One of the things you can do to make the signal go out further is to make the donut larger in diameter, because there is only a certain amount of power to work with—five watts input and four watts output. The donut can be squashed flatter and made thinner with the result that it would have a larger diameter.

One way of doing this is with what is called a ground *plane.* A ground plane is any object that the signal would treat like a solid metal surface under the vertical antenna. It can be four pieces of wire connected to ground radiating in four different directions at the base of the antenna. The signal is pushed up off the ground, flattening the bottom of the donut and causing energy to radiate in a larger circle around the antenna.

A modification of the ground-plane antenna is called the *drooping ground plane* antenna in which the four ground plane wires tilt downward at about a thirty-degree angle. To squash the top of the donut down here, you use a device called a *top hat.* This goes at the top of the antenna. It can be either a disk of a fixed diameter or what looks to the signal like a disk of a fixed diameter. It can be a metal ring with four rods connecting to the antenna. The donut is squashed down from the top and when you are squeezing it up from the bottom and down from the top like that, you get further signal distance because it has no place to go but out.

All right! We've taken a whip antenna and added a plane to it. It's now a drooping ground plane with a top hat on it. Not too attractive, I'm afraid. Many people said they wouldn't like to use a directional antenna on their house. They would rather use an omni-directional antenna, such as the standard vertical whip and that would be perfectly all right if they could manage to mount it up in such a way that it wouldn't be interfered with by the structure of the house.

Other people felt that while they wanted a good antenna, they didn't want to put it where it could

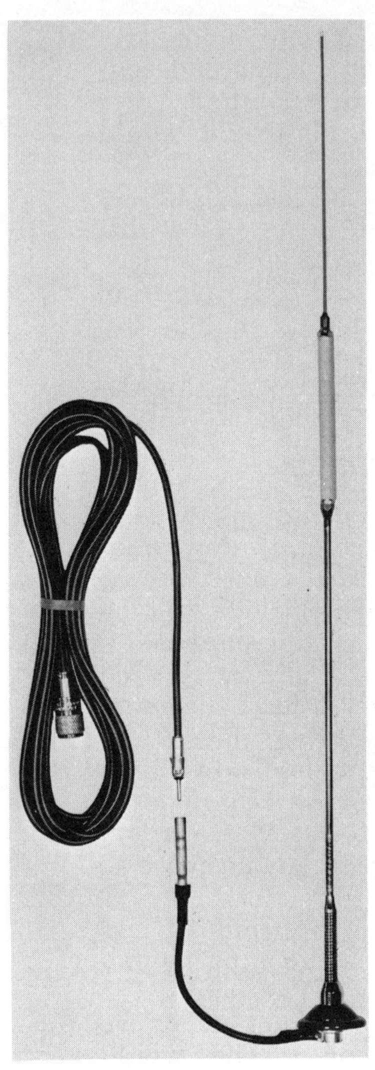

Roof Mount 25"
"High Efficiency"
CB Antenna
Model RTS-27L

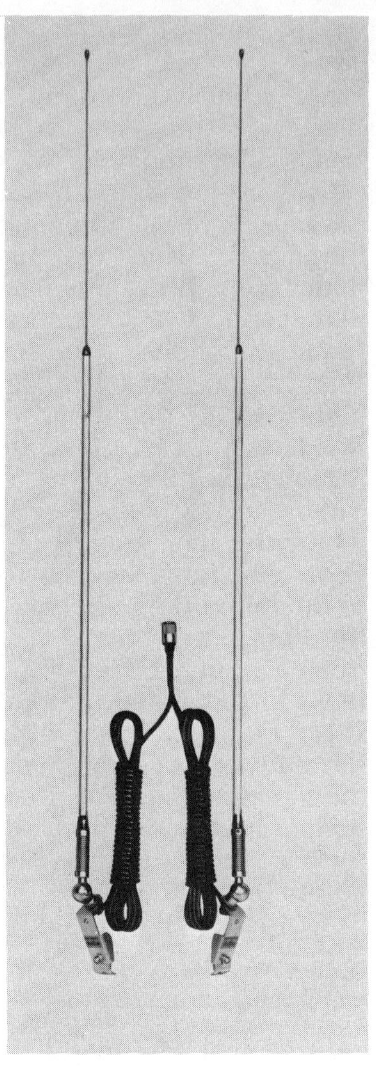

Dual 42" Fiberglass Antenna
System with Rain Gutter Mounts
Model DFG

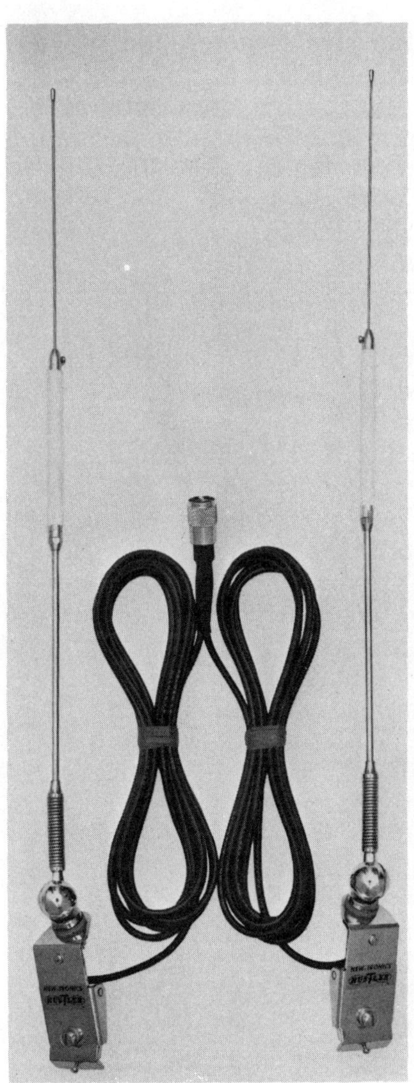

"High Efficiency" 25" Dual
Rain Gutter Mount Antenna
System. Model DTG

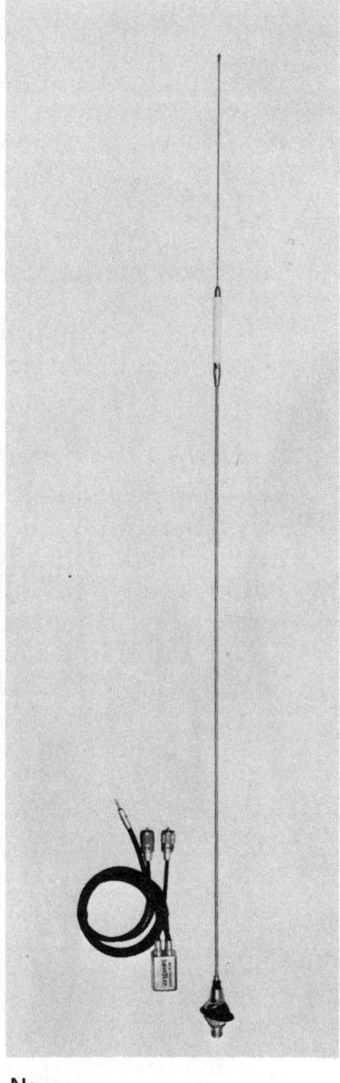

New - - -
   Dual Purpose CB
AM/FM Antenna Replaces
Auto Antenna or for New
Installation. Model FGB-27M

be affected by the wind and weather. So where can you put it? In the house? Sure! Unless, of course, the roof of the house is covered with metal or some metalic or foil type of insulation is inside the attic walls. The fact is that the C.B. signal will never even see the roof of the house anyway and will appear as if it were outdoors.

When considering an antenna, buy the best one your own particular wallet can afford. Some people go out and spend a few hundred dollars on a C.B. unit, but then go out and try to purchase the cheapest antenna possible. They may as well use coat hangers, because buying a cheaper antenna will deprive you of the best performance your transmitter can bring. Often, whenever you think you bought yourself a bargain, you only end up by going out and buying a dozen or so more bargains. So don't fool yourself. If you've got the best transmitter in town but also have the cheapest antenna, forget it!

One thing to be concerned with when selecting an antenna is considering what transmission line you are going to use. The transmission line is that which connects the antenna to the transceiver. If your transceiver has two terminal screws on the back marked antenna, then it is a balanced output. To get the signal from the transceiver up to the antenna you would use either a fifty- or three-hundred-ohm flat transmission line. This looks like a television antenna wire. It is two parallel wires that are encased in a flat plastic package.

So, if those two terminal screws are what you have, and you are planning on using the flat transmission line, you will then use a balanced antenna, a dipole type, or a *yagi* (a beam type antenna). You cannot connect a three-hundred-ohm transmission line directly to a vertical whip. There is no provision for it. This would be an awful mismatch because the antenna requires a fifty-two-ohm feed

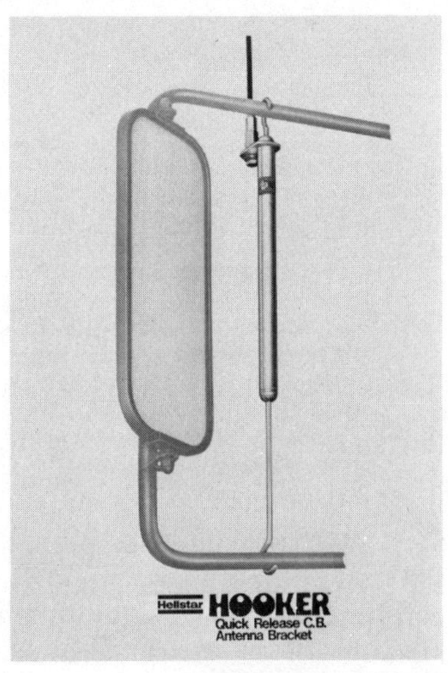

COURTESY HELLSTAR

and the transmission line is three-hundred-ohm. No, you cannot connect a flat transmission line directly to a vertical polarized whip antenna.

If you have a coaxial antenna on the back of your unit, there is no way to connect that flat transmission line. You must use coaxial cable which is terminated with a proper matching plug. The two most common types of plugs used today are the BNC and the Pl. The first of the two uses small fifty-two-ohm transmission line, about a quarter to five-sixteenths in diameter. The larger one uses one-half-inch transmission line. But in either case, the transmission line consists of a center conducter which is surrounded by a flexible plastic jacket. This, in turn, is surrounded by a braided conductor called a *shield*. A rubber covering protects the shield. The two conducting wires are coaxial; they have the same axis. This measures fifty-two ohm. There are two types and they are RG/52-U and RG/52.

The Tarantula, Model CT-2
The newest member of the Mosley "Spider Family"
of mobile antennas. Designed for truck or
recreational type vehicle, the CT-2 features a
mirror or luggage rack clamp-mount for quick,
easy, no drill mounting.

A simple whip adjustment allows antenna peaking
at desired frequency between 26.9 and 27.3 MHz.
Power Rating; 500 watts (input to the final amplifier.)
SWR at resonance; 1.5 to 1 or better.
Transmission line; 12 ft. 52 Ohm (RG-8/U) coax supplied.
Assembled height; Maximum 4½ ft.
Assembled weight; Approximately 3/4 lb.
Shipping weight; Approximately 1 3/4 lb.

Mosley *Electronics Inc.*

The 52-U is made with a center conductor consisting of several fine wires that are twisted
together. The 52 is a single-center conductor of
the same diameter as the twisted finer conductors.
The difference is that with the RG/52-U cable you
get much more flexibility. It bends and gives a lot
more than the one with the solid center core.

So, if you have two terminal screws on the back
of your transceiver, you are going to use a flat
three-hundred-ohm transmission line. And if you
have a coaxial connector you are going to use coaxial cable as a transmission line. Both of these will
get your signal up to the antenna without any
trouble.

Sometimes you could run into a problem when
you have a coaxial connector and a three-hundred-
ohm antenna or a three-hundred-ohm connector on
the back with a whip antenna of fifty ohm. As you
can see, there is a mismatch. It has to be compensated for and you do this with a device called a
balun.

A balun consists of a small loop of coaxial cable
cut to a specific length and connected in such a
way that it will convert three-hundred ohm of

transmission line at the antenna to fifty-two ohm, and fifty-two ohm of transmission line to three-hundred ohm.

Generally, people will go out and purchase a piece of wire and make their own baluns. This is fine. Still, others prefer to go out and buy a *matching transformer*. This is a device whereby you could hook up at the antenna and match anything to anything and come through with a perfectly good, clear signal.

Now, one of the most important factors to take into consideration when you are selecting your antenna is what kind of a connector is going to be used at the output of your transceiver. If it is a coaxial type, and most of them usually are, well then, your antenna should be a fifty-two ohm or a vertical whip.

However, if it only has terminal screws, you would probably do better with a dipole antenna— maybe even a folded dipole or a yagi.

Now, many of the units already have a matching transformer and a balun built right into the set.

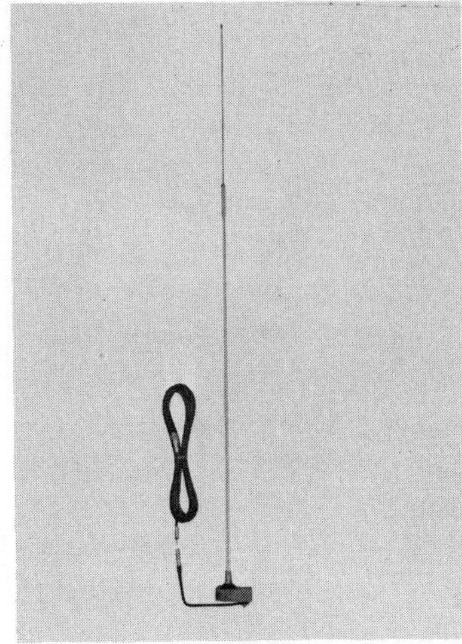

Trunk lip mount, 48" center loaded CB mobile antenna. Model TLA-27L

Then you will be offered two types of antenna connectors on the back skirt of the chassis.

You will have a pair of terminal screws if you are going to use three-hundred-ohm and you will also have a coaxial connector if you intend to use fifty-two ohm. Here, you've got the option to do whatever you wish.

There is still another factor that should be considered when selecting your antenna, and that is it is going to be mounted. And wherever you want it put, whether it is going on the car or the house, think about the kind of mounting you want to use. The previous chapter on installation will be of assistance to you in determining this.

Here is yet another point to be concerned with: Suppose you do not live close to the city? Maybe you are a farmer and you reside in a valley in hilly country. The hills might shorten the distance of the range of your C.B. radio.

If your house is within a few hundred feet of the hill, you could mount your antenna there. You see, a base antenna can be mounted on top of or at one end of a house. When a tall building or some other kind of tall structure is adjacent to the house, you could mount your antenna on top of that, if you

94

are able to get the permission to do so. In this case, a low-loss coaxial cable should be used.

If you need as much as twelve-hundred feet of coaxial cable, then you can use a seven/eighth inch foam cable and you can support this with wooden poles or buy it and cover it with some kind of protection, such as polyethylene. This cable comes in maximum lengths of three-hundred feet, so you will have to use splicing connectors. Loss of transmitting power can be made up by using a gain-type of antenna.

You see, an antenna can act as an amplifier. It provides what is called "power gain." An omnidirectional gain-type of antenna compresses the radiated energy toward the earth so that less escapes. This is very good for jumps of fifty feet where you would have a twenty percent power loss.

Perhaps you would want to consider a co-phased antenna which is a bidirectional type affair. This is the use of two antennas for one radio. You can co-phase any type of antenna, thus bringing about double signal strength. A cable goes from the back of the car radio where the wire splits into two separate pieces and these go to each antenna, the energy radiating toward the front and back or to the sides.

There is still another type of antenna—a beam type affair, commonly called "big mama." A beam directs radio energy generally in one direction. Here! You can make your own illustration.

Turn on a lamp. Then shine a flashlight on a wall which is illuminated by the lamp. You will notice that the flashlight beam is brighter than the lamp even though the lamp is radiating much more light. This is because the flashlight beam is concentrated in one direction.

So, a given signal strength into the antenna will go farther in the beam's "forward side." However, there will be less signal strength off the sides and back. Now when receiving, a beam will be more sensitive at the front, and less sensitive off the sides and back.

The comparison is an omni-directional antenna, or antenna that radiates equally well in all directions. It is the equivalent of the lamp used in the illustration. Okay, let's put the lamp away for now.

So, when you finally make your choice about which antenna you wish to purchase, remember to buy the best one you can afford. Make sure you do not miss out on the best performance your C.B. radio can bring.

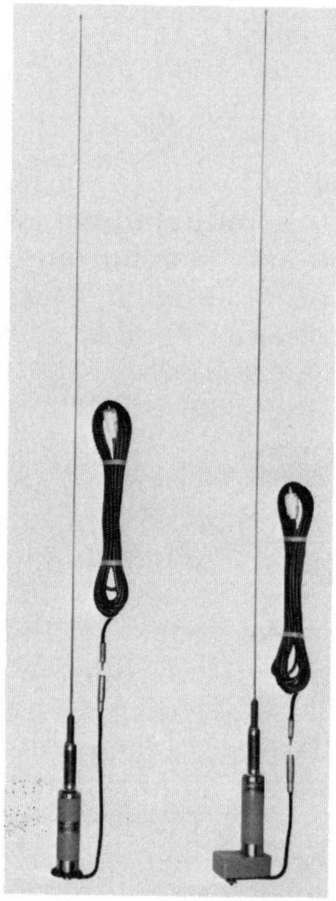

Roof Mount 48" base
loaded CB antenna
Model XBL-3

Trunk lip mount 48"
base loaded CB antenna
Model XBLT-3

COURTESY NEW-TRONICS CORPORATION

CHAPTER 7

ACCESSORIES

Microphones

The Carbon Microphone. Carbon microphones were used in the early days of broadcasting. They were the only types available. The quality was pretty awful. They produced hissing noises when they were being used.

Although Citizens Band radio is not really concerned with broadcasting, it is certainly interested in the advancement of clear communications. So let us, for a moment, go back and investigate that old carbon mike.

Carbon microphones were made of a thin metal diaphragm, which was a disk of about one and a half to two inches in diameter. An absorbent cotton ring was cemented to the back of it. This was called a *washer* and inside of it were carbon granules. A plunger fit at the other end. This entire unit was encased in a metal or plastic sleeve, or perhaps, a holder with holes punched in front of it so that your voice would hit the diaphragm, causing it to flex back and forth.

When the diaphragm went back, it squeezed the carbon granules together. When it went forward, it loosened them to a slight degree. Then, a small voltage of approximately one and a half to three volts of direct current was passed from the diaphragm to the plunger disk at the rear. Voltage was forced to go through the carbon granules.

When the carbon granules were squeezed together, they came to a lower resistance and more voltage passed through. So talking at this diaphragm caused resistance through the microphone which was fed to the audio amplifier, then to the modulator stage of the transmitter. So that was how sound was sent out over the airwaves.

Although carbon microphones seemed highly advanced, there were flaws in the technology of these instruments. One fly in the ointment was that after a short while, the carbon grindings would tend to pack. They wouldn't relax when the diaphragm did. To make the microphone work better, the granules had to be loosened. To do this you had to bang it against your fist or on a table.

This microphone was great for the army because it was rugged. If they were out drilling, or whatever, and they were surrounded by rough conditions, the microphone would thrive on its being banged against things or dropped repeatedly. Though you would think all this hard treatment would ruin the instrument, it actually did it good because the carbon granules were constantly being knocked around and loosened, therefore, enabling better and clearer communications.

The Crystal Microphone. The crystal microphone eliminated the need for the old carbon microphone. It used a piezoelectric crystal which, when flexed, gave off a small voltage. So, one end of the crystal element was attached to the back and the other end to the diaphragm. When a voice hit the diaphragm, it caused the crystal to flex and give off a small voltage which could then be amplified and go through the modulator stage of the transceiver.

But the crystal microphone had its troubles, too. It was an extremely delicate instrument. The crystals could be broken very easily. Once the crystal broke, the microphone element was no good anymore.

The Ceramic Microphone. The ceramic microphone came along and replaced the crystal microphone. In this microphone was a ceramic wafer of man-made material. It was a little more rugged, and also a little less expensive to produce. The ceramic could be poured into molds and baked

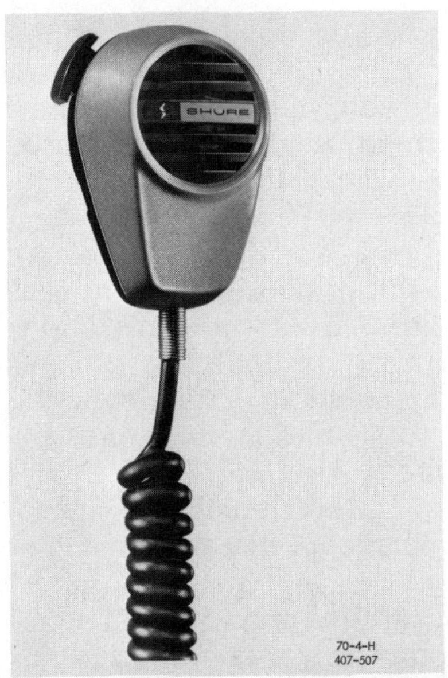

7-1 Typical modern microphone features such conveniences as push-to-talk operation and a retractile coiled cord to keep the mike from fouling operating car equipment.

COURTESY SHURE BROS., INC.

7-2 Added operator convenience is afforded in this desk mike that can be raised or lowered to suit the operator's individual height.

COURTESY SHURE BROS., INC.

whereas crystals had to be very finely cut and polished.

However, the ceramic microphone also proved to be a very delicate instrument and it operated much in the same way as did the crystal microphone.

A phonograph cartridge is very much akin to a microphone. The only difference is that the needle (stylus) rides in the grooves of the record and is attached to the diaphragm, so that the grooves cause the diaphragm to move up and down the same way as your voice does in a microphone.

The Dynamic or Magnetic Microphone. The dynamic microphone (also called a magnetic microphone) uses no electricity except that which is generated by itself in its coil.

If you push a magnet rapidly in and out of a coil of wire, an electric current is generated. The intensity of the electric flow generated will be determined by three factors: how rapidly the magnet is pushed back and forth; how strong the magnet is, and how many turns there are in the coil.

By changing any of these factors, the voltage also changes. A coil of wire is attached to a small magnet which is attached to the back of the diaphragm and connected to the coil through proper cables, which is connected to the amplifier or the modulator or the transceiver. When you talk into the diaphragm, the magnet moves and so does the coil. The louder you speak the faster they move. That is how this particular microphone works.

The Electronic Condenser Microphone. This common type of microphone and is also known as the ECM. A condenser, also called a capacitor, consists of two metal plates with a small voltage applied through them. Capacitance is the build-up of an electric charge between two dielectric plates. The charge is measured in farads or micro-farads, and this is determined by the spacing between the two plates. A farad is the unit of electrical capacitance. There are one thousand micro-farads in a farad. The dielectric is the area between the two plates. On fixed capacitors the dielectric can be air, paper, plastic compositions or oil. For example, the tuning knob on your radio is attached to a capacitor and that capacitor uses air as a dielectric.

Anyway, the two plates are fixed. There is the *rotor* plate and the *stator* plate. The rotor plate turns when you turn the knob, and meshes with the fixed plates. If you take them out, there is less capacitance. But if you put them back in, well, then there is more. Capacitance is the build-up of electric charge between the dielectric plates.

In the electronic condenser microphone the two are called the plunger plate and the diaphragm. You just apply a small direct current voltage, and then by speaking at the diaphragm, you move the plates closer together and further apart.

This is a very sensitive kind of microphone. This is the type of microphone you ought to use for your communications, as it is plain to see. When you buy a C.B. radio, you won't be able to use it without a microphone, so the manufacturers are obligated to sell you one with your transceiver.

Now, the manufacturer will usually enclose some literature with your unit, suggesting what kind of a microphone you should purchase. It would be a good idea to follow his recommendation for your own sake.

Perhaps you might want to change the microphone you have. For example, some of the less expensive transceivers come equipped with push-to-talk function. If you buy a transceiver that doesn't come with a push-to-talk microphone and your literature states that your set can use this type of mike, you can do so.

However, it would be best to buy a set that already comes with a push-to-talk function, because it would be a big job to have to undress the chassis.

Ceramic Microphone. The Mura Corporation has introduced the DX-116 new ceramic microphone. It is equipped with a push-to-talk switch wired to accommodate electronic or relay switching. It is a gain-controlled microphone and has a newly designed transistor amplifier powered by a standard nine-volt battery capable of providing sufficient output to 100% modulate most transmitters. This microphone provides C.B. users with the facility to get the extra performance often demanded in numerous communications applications. A slide-type gain control is mounted on the front of the microphone case to adjust the output level from the high gain amplifier. It sells for $19.94.

Lightning Arrester Accessory. An interesting accessory is one called a "Blitz Bug" or lightning arrester. It resembles a coaxial connector, about three inches long with a screw on it. You would run a piece of wire from that screw to the nearest ground, such as a radiator, if you have one handy, and make a good solid connection at both the radiator and the blitz bug. A coldwater pipe can be used instead of the radiator.

The blitz bug comes with a specially pointed screw which you screw down very tight to make a positive bond. Then you connect the blitz bug to your antenna terminal on your transceiver. Next, connect the other end of the blitz bug to the antenna. In ordinary circumstances, it permits a signal to go through easily in both directions. And when the antenna is struck by lightning, the heavy bolt of voltage, instead of it going into your transceiver and destroying it, is passed to ground harmlessly.

Field-Strength Meter. Another accessory that might interest you is the field strength meter. It is inexpensive and consists of a meter, a rectifier and a short wire stub antenna. It has no batteries, is

not connected to anything, but sits on your table-
top or whatever.

But when you are busy talking, and your transmitter is on, you could see the needle bounce with
your voice. This device simply tells you that your
transmitter is working. After you have used the
field-strength meter for awhile, you'll get to understand where the needle should bounce. One day
you could be talking and you could notice that the
needle is hardly moving. You'll know at once
when something is wrong.

7-3 Portable field strength meter samples the signal from your transmitter and provides a relative field indication of signal strength.
COURTESY MURA CORPORATION

SWR Meter. Some of the more expensive units
have this accessory. It measures power generated
by your transmiter and reflected power which is
power reflected back to the transmitter by the antenna. The more reflected power, then the less
power is absorbed and radiated by your antenna.

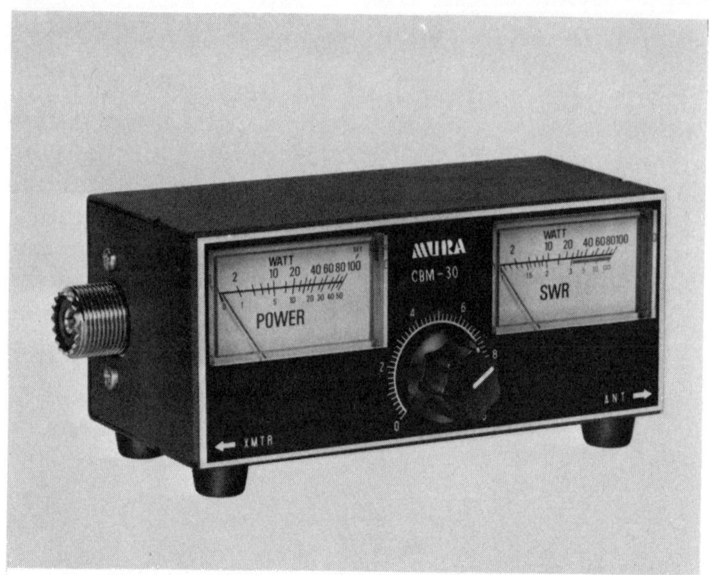

7-4 SWR, the "Standing Wave Ratio" can inhibit the proper flow of signals from your transmitter to your antenna. The SWR meter sees to it that your antenna feedline is pruned to the proper length for optimum signal transmission.

Earphones. Sometimes they are included with your unit. The value of earphones should not be underrated. Earphones should not be thrown away. You'd be surprised at how often they may be used. For instance, suppose the other members of your family want to watch television in the evening, or just chat or listen to music. It would be unfair of you to curtail their activities and it would be just as unfair if they were to stop you from yours. The earphones could really solve this problem.

When you plug the earphones in, there is a cut-off circuit that disables the loudspeaker so that the sound appears only in the earphones.

The earphones supplied with your set are usually the inexpensive kind and are not best suited to do the job. If you go out to buy a pair of

104

earphones, the kind you would buy would depend on the kind that came with your transceiver. Usually, with a transceiver, an instruction booklet is included that contains a schematic diagram. But you do not have to know how to read this diagram. Just look over the schematic to see where the loudspeaker is located and how many ohms the loudspeaker is. If it is an A-ohm loudspeaker, then get an A-ohm earphone set. If it is a sixteen-ohm loudspeaker, then get a sixteen-ohm earphone set.

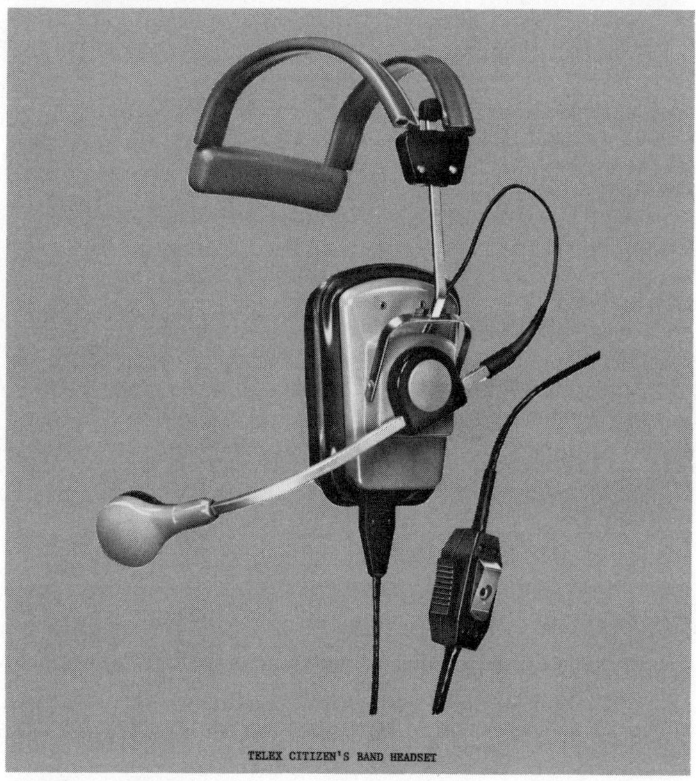

7-5 Combination headset consists of an earphone, a boom mike that can be positioned near your mouth, and a cable-mounted push-to-talk switch. It's the ultimate in comfort and operating convenience.

Scanner. Tennelec has come out with a portable scanner that lets you hear the action on police, fire, weather, marine, rescue, business, and other frequencies, as it happens, regardless of where you are. Because of its small size and light weight, you can carry it with you from room to room at home or you can use it in your car, the office, or anywhere you would want to go. With this scanner you can monitor up to four channels at a fast scanning rate of twenty-five channels per second. A two-second delay occurs after a message, allowing enough time for response. You can change the selections by changing crystals.

This scanner costs $99.95 and comes with dual filters, earphone jack, long-life LED indicators, a fully collapsible multi-position antenna, an optional power jack for 9V DC-AC power converter and recharges. It can use standard crystals.

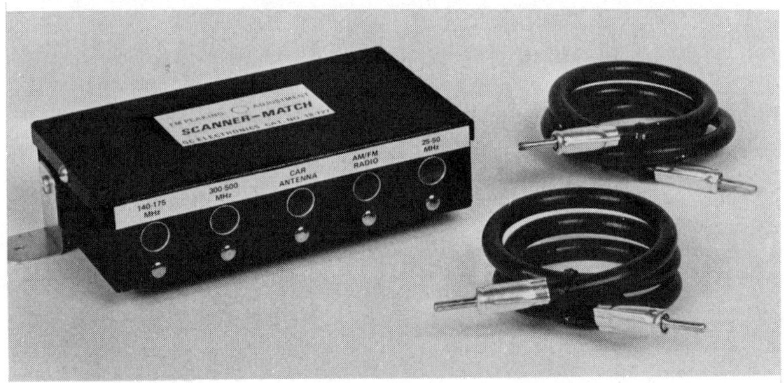

7-6 Whenever you introduce a new element in your system, you had better be sure that it will not create a power-losing mismatch. A Scanner-Match, such as this, can be mounted out of the way, for it carries no operating controls, assures proper match to other system components. COURTESY G. C. ELECTRONICS

Clip Board. It is approximately eight inches long and about four to six inches wide. It straps to your upper thigh just above the knee. If you are a woman, it certainly would be more comfortable to put it on over pants. It is like a desk on your lap and will not move.

Some clip boards come equipped with batteries and a small light to illuminate the writing surface. There is even a little slot on the side to hold a pen or pencil. Pilots use standard pads (six by eight) which come in very handy, especially when you are traveling and you want to jot down some notes. These accessories usually cost about sixteen dollars.

There are many other interesting accessories, such as an **AC power supply.** This converts mobile radios into desk-top base station operations. **Portable power packs** convert mobile units into handy portables with built-in rechargeable batteries. **External speakers** can add increased clarity under noisy conditions. An **in-converter** permits use of any negative ground mobile radio with positive ground systems. It can be used to operate radio from six volt systems. A **transceiver tester** can check the performance. It reads power output in watts and modulation in percentage; has built-in RF and audio generators, and crystal acitivity checker; SWR meter and a built-in antenna for field strength meter readings.

Another outstanding accessory is one called a **television interference filter.** You see, the frequency of a C.B. transmitter is about twenty-seven megahertz (abbreviated mhz, twenty-seven million cycles per second). The frequency of the lowest television channel—channel two—is between fifty-four and sixty mhz. Television signals are very wide—six-hundred C.B. channels could fit into the space occupied by a single television channel.

Sometimes a C.B. transmitter emits harmonic signals, or signals other than those of the frequency they are supposed to be. For example, the first harmonic of twenty-seven mhz is fifty-four mhz, right smack on channel two.

7-7 Maybe YOU like C.B. radio, but chances are that your neighbors won't—not if you start calling on the air and break with Marcus Welby! It's a simple matter to install a TVI (for Television Interference) filter and save all the woe.

COURTESY G. C. ELECTRONICS

To reduce the probability of interference, every C.B. base station should have a low pass filter right in the line between the transmitter and the antenna which passes low frequency signals but blocks higher frequency signals.

If television sets in the area are equipped with high pass filters as well, chances of interference would be even less, as they usually pass the higher frequency television signals but block the lower frequency C.B. signals.

CHAPTER 8

HOW TO USE CITIZENS BAND RADIO

As we noted in previous chapters, Citizens Band radio is for short-range communication. It can be used by any citizen over the age of eighteen.

C.B. radio takes a certain amount of "getting-used-to," the same way you would have to adjust to looking at your new color television set instead of the old black and white picture tube. However, in the end, you will wonder how you ever got along without it as long as you did.

When you purchase your rig, you have to fill out the enclosed 505 form and send it to the Federal Communications Commission along with a four-dollar fee. There is a long wait for the license to be issued, but you should begin talking over the air only after the license has been issued.

Because the band is usually crowded, courtesy among the C.B. radio users is always practiced. If you want to get on the air and you find somebody else is talking—and there always is—you adopt the following procedure: Suppose you want to get on channel nineteen, and you find another party using it. You would listen for a moment, just in case the person might be nearly finished talking. If he stays on the air you would say, "Break, break, 19." Then that person would say, "Go ahead breaker." Now you can give your handle and your call sign.

Name and Call Sign. Though most C.B. users today give just their handle, it is actually an illegal practice. The FCC assigns a call sign to every C.B. user. It should always be used. This rule is so widely broken that many people don't even know what the rule is. Well, it is in FCC Rules 95.95. Look it up for yourself to make sure. You will find that you are, indeed, supposed to use your call sign.

You can use any handle you want. Someone I know uses the moniker Pacemaker. Know why? Right! That person has a pacemaker in his body. Suppose you write for a living. You can call yourself the "writer." Maybe you are a speed car racer and you call your racing automobile Little Rocket. You can use that name, too. There is no end to the handles you can come up with. It's fun.

Prohibitions and Restrictions. You should limit yourself to about five minutes per stretch on the air. Say what you have to and then give the other person a chance.

One thing you must not do over the C.B. radio and that is use foul language. The FCC does listen in and there could be a heavy penalty if you are caught.

You cannot use the radio for any purpose contrary to federal, state or local law.

You must not whistle, sing or use sound effects to attract attention or just for the fun of it, or for any type of amusement purposes.

Do not use your rig for relaying messages on behalf of another person unless there is an

110

emergency and you are trying to assist a motorist. You cannot use your radio to say "Mayday" or other international distress signals except where a human life is in grave danger. You cannot advertise or solicit the sale of goods or services, and you must not transmit in a code other than in plain language.

You cannot use your rig to transmit over a distance of more than one-hundred and fifty miles. Your C.B. radio must not be used for relaying messages about technical performance of the radio

equipment, except if it is necessary to maintain the specific communication.

You also cannot use your radio for direct transmission to the public, in the manner of a public address system.

Citizens Band radio must not be used for pay, nor for communications between unlicensed stations, or stations that are not licensed in the Citizens Radio Service, such as hams.

You cannot use your unit to act as a broadcasting station, which means transmitting of messages not directed to a specific station.

You are prohibited from using your C.B. unit for relaying messages that are to be broadcast over regular AM, FM, and television stations. For instance, let us say that a C.B. operator is on the scene. He transmits an account of an event which is then picked up and used by a news reporter for a broadcast station. This is just not allowed by the Federal Communications Commission.

You must never interfere intentionally with another station or transmit communications that are not necessary.

For example, a C.B. operator transmits a message that a motorist is in need of assistance and as part of that message also mentions the fact that he bought his Citizens Band radio at a particular store.

Servicing Your Rig. If your rig malfunctions in any way, take it to a reputable dealer. Describe the problem you seem to be having and let them fix it for you. You are far better off using such professional help.

But you might say, "Wait a minute! Service costs money and I know a kid who lives down the block who knows all about fixing C.B. radios real cheap."

Okay! Here are the "facts of life" about having your Citizens Band radio repaired. Service does cost money. In truth, it could cost as much as twenty dollars an hour to have a professional mechanic work on your rig. But you must remember that a quality shop with a top repairman has a well-equipped lab. He can find the problem and fix it quickly and competently.

The high-school kid or Cheap Charlie over on Vine Street who works for only seven dollars an hour may not have even as much as an accurate voltmeter going for him. He's liable to end up giving your radio much more trouble than you had in the first place.

You see, stores that provide quality service cannot afford to give discounts and stores that discount cannot afford to give quality service. The store pays so much for the radio wholesale. It has to pay for its rent and utilities, and then there is the advertising, not to mention the salaries for salespeople. If it pays for quality service, too, then there is just no room for discounts and still make the profit that makes the wheels go round.

Summary of Operating Procedures. Okay! So let's go back and recap how to communicate over your C.B. radio, keeping in mind that channel nine is used for emergencies only.

1. To get on the air, announce the call sign of the person you want to talk with.

2. To answer a call, announce your own call sign and also the call sign of the party who is calling you.

3. When you want to conclude a conversation, you announce your call sign and the call sign of the party you are communicating with which is followed by the words "off and clear."

4. The length of your conversation should be no more than five minutes. If you have to talk for more than five minutes, wait five minutes before getting back on the air.

5. Using a handle can be fun. You may choose any one that appeals to you. When you announce yourself over the air with your handle, be sure you also give your call sign as part of your recognition.

6. If you want to use the channel and you find it is being used, wait until it is clear. Otherwise, if you have an urgent need to communicate, say "break-break" and state your reasons for wanting to break in on the other party's conversation.

Glossary of Terms and Jargon Used by C.B. Operators. With your new Citizens Band radio all set to go, your license in hand, here is a list of the lingo that C.B. operators use, and what these terms mean.

ADVERTISING—A marked police car that has its lights turned on.

BACK DOOR—Last vehicle in a string of three or more, all in contact with each other. (Used mainly by truckers, and refers to a truck)

BEAR—Policeman.

BEAR CAVE—Police station or post on highway.

BEAT THE BUSHES—"Front door" (lead vehicle) looks for Smokey (see below) by going fast enough to draw him out of hiding. It is the same as "Shake the leaves."

BODACIOUS—Good signal; clear transmission.

BREAK ONE-OH (also "Break 10")—I want to talk (on Channel 10).

BUSHELS—One-half-ton load; a twenty-ton load would be forty bushels.

CAMERA—Police radar unit.

CATCH YOU ON THE OLD FLIP/FLOP—Catch you on the radio on a return trip.

CHECK THE SEATCOVERS—Watch out for a female driver with her skirt pulled up.

CHICKEN COOP—Highway truck-weigh station.

CLEAN—No Smokies (Smokey means a policeman) around.

COMIC BOOKS—Truckdrivers' log sheets or log books.

COTTON PICKER—Cotton picker (instead of four-letter words on the air).

COUNTY MOUNTY—County sheriff or highway patrol.

EARS—Antennas or radios. (See also "Smokey with Ears.")

EATUM-UP—Roadside restaurant.

EIGHTEEN WHEELER—Any semi-tractor truck with any number of wheels.

FAT LOAD—Overload, more weight than local state law allows.

FEED THE BEARS—Collect a ticket from Smokey.

FIVE-FIVE—55, the legal limit in most places.

FOUR—Abbreviation of "10-4," meaning "OK."

FOUR TEN—10-4, emphatically.

FOUR WHEELER—Passenger car.

FRONT DOOR—First vehicle (truck) in string of three or more trucks in radio contact.

GRASS—Side of the road or median strip.

GREEN STAMPS—Dollars.

GREEN STAMP ROAD—Tollway.

HAMMER—Accelerator.

HAMMER DOWN—Highballing; driving fast.

HANDLE—Slang names often used by truckdrivers instead of their call letters.

IN THE GRASS—Parked or pulled over on the median strip.

KEEP YOUR NOSE BETWEEN THE DITCHES AND SMOKEY OUT OF YOUR BRITCHES—Drive safely and look out for speed traps and speeding fines.

KEEP THE GREASY SIDE DOWN AND THE SHINY SIDE UP—Drive safely.

KENOSHA CADILLAC—Any car made by AMC.

LET THE CHANNEL ROLL—Let others break in and use the Channel.

MERCY—Oh, wow!

NEGATORY—No. Negative reply.

ON THE MOVE—Driving, moving.

ON THE SIDE—Parked or pulled over on the shoulder.

OTHER HALF—Wife or husband.

PLAIN WRAPPER—Police car with no markings; an unmarked car.

PICTURE TAKER (same as "Camera")—a police radar unit.

PICKUM-UP—Light truck; pickup truck.

POUNDS—Number on S-meter (S-3 is three pounds, for example.)

PREGNANT ROLLER SKATE—Volkswagen.

PUT THE GOOD NUMBERS ON YOU or THREES AND EIGHTS—best regards.

RAKE THE LEAVES—Back door or last vehicle in string, bringing up the rear.
RATCHET JAW—Nonstop talker.
REST-UM UP—Roadside rest area.
RIG—Citizens Band radio; tractor
ROCKING CHAIR—Vehicle that is between the front door and back door in a string of vehicles.
ROGER ROLLERSKATE—Passenger car going more than 20 mph over the limit.
ROLLER SKATE—Small car.
SEATCOVERS—Occupants of passenger car, usually with attractive females.
SHAKE THE LEAVES—Act as lead vehicle to decoy any Smokies out of hiding (same as "Beat the Bushes").
SIX WHEELER—Passenger car pulling a trailer.
SMOKEY—The police.
SMOKEY ON FOUR LEGS—Mounted police (used in New York City and Chicago)
SMOKEY THE BEAR—State Police Patrol (with or without a Smokey the Bear hat).
SMOKEY WITH EARS—Police listening on C.B.
STACK THEM EIGHTS—Best regards.
SWEEPING LEAVES—Bringing up the rear. (Also see "Back Door," and "Raking the Leaves.")
THIRTY THREE—10-33; This is an emergency.
THREES ON YOU—Best regards.
THREES AND EIGHTS—Lots of best regards.
TIJUANA TAXI—Well marked police car.
TRAIN STATION—Traffic court that fines everybody.
TWO WHEELER—Motorbike; motorcycle.
TWO WAY RADAR—Radar used from moving police car.
WALL TO WALL—Peg full-scale on S-meter.
WALL TO WALL BEARS—High concentration of police with strict enforcement, traps, etc.
WE GONE—Stopping our sending; will listen.

WRAPPER—Color; "Blue Wrapper" is a blue car, usually refers to an unmarked police car.
XYL (stands for ex-young lady)—Wife.

Glossary of Strictly Truckers' Terms. Some of these you may find interesting and amusing:

ANCHORED MODULATOR—Base station operator.
BAO BAB—Wide load.
BIG ORANGE—Snyder truck.
BLINKIN WINKIN—School bus.
BLOOD BOX—Ambulance.
BOUNCE AROUND—Next trip through.
BUBBLE TROUBLE—Tire problem.
CHICKEN CHOKER—Poultry truck.
CIRCUS WAGON—Monofort truck.
CORNFLAKE MACHINE—Consolidated freight.
DEAD PEDAL—Slow moving vehicle.
DEISEL DIGIT—Channel 15.
DRAGGIN' WAGON—Wrecker.
FLAG WAVER—Highway worker.
GOLDIE LOCKS—Mobile business women.
GOOD NUMBERS—73rds and 88s.
HALL HAUL—Halls motor.
HALLOWEEN MACHINE—Cooper-Jarrett truck.
HOT PANTS—Smoke or a fire.
JAW JACKING—Conversation.
KODIAK WITH A KODAK—Police with camera.
LIMP LINE—Rigging loose; shifted load.
MOBILE EYEBALL—Check operator's rig while moving.
MOBILE PARKING LOT—Automobile transport.
MONSTER LANE—Inside lane.
ON THE PEG—Legal limit.
PORTABLE ROAD BLOCK—McClean truck.
PUT THE HAMMER DOWN—Run.
SAFER SHAFFER—Shaffer trucking.
SALT SHAKER—Cinder truck.

SMOKEY DOZING—Police in car, stopped.
SMOKEY ON THE GROUND—Police out of car.
SMOKEY ON THE RUBBER—Police moving.
THERMOS BOTTLE—Tank truck.
WALLACE LANE—Middle lane of a three-lane highway.

On Being Clearly Understood. Although proper maintenance and operation of your C.B. equipment is very important, it is equally important to make yourself clearly understood over the air. In order to reduce on-the-air time, because you do share the channels with others, and to insure speedy and accurate transmission of messages, the police radio operators have developed a working code to enhance efficiency.

APCO, Associated Public-safety Communications Officers, Inc., recommends the APCO Brevity Code of "Code-10" Signals to be used by everyone. Often misunderstandings in radio communications are due to the improper pronunciation of unusual names such as for persons and places. Whenever there might be the possibility of an error, spell the names and, in turn, request that the names be spelled out when you are receiving. REACT, Road Emergency Associated Citizens Teams, has recommended the use of a phonetic alphabet to aid in improving signal readability and intelligence (see below).

You will note that Section 95.83 (16) of the FCC Rules and Regulations warns that codes or other recognized operating signals may be used only if a copy of the code and its meaning is kept on file in the station records and made available to any FCC representative on demand.

International Phonetic Alphabet

(A)	ALPHA	(N)	NO-VEMBER
(B)	BRAVO	(O)	OSCAR
(C)	CHARLIE	(P)	PAPA
(D)	DELTA	(Q)	QUEBEC
(E)	ECHO	(R)	ROMEO
(F)	FOXTROT	(S)	SIERRA
(G)	GOLF	(T)	TANGO
(H)	HOTEL	(U)	UNIFORM
(I)	INDIA	(V)	VICTOR
(J)	JULIETTE	(W)	WHISKEY
(K)	KILO	(X)	X-RAY
(L)	LIMA	(Y)	YANKEE
(M)	MIKE	(Z)	ZULU

Phonetics should always be used as: A—Alpha, F—Foxtrot, O—Oscar, L—Lima, and so on. Never use the expression "A as in Alpha," or "F as in Foxtrot," or "L as in Lima," etc.

The Brevity Code

10-1   Signal weak—change location
10-2   Signal good—continue
10-3   Radio silence (stop transmitting)
10-4   Affirmative-acknowledgement (O.K.)
10-5   Relay (to)
10-6   Busy
10-7   Out of service (at
10-8   In service (at
10-9   Say again-repeat
10-10 Negative (denied)
10-11 On duty (who is on duty?)
10-12 Stand by (stop)
10-13 Existing conditions
10-14 Message (information)
10-15 Message delivered
10-16 Reply to message
10-17 Enroute to assignment

10-18 Urgent
10-19 In contact with or are you in contact with . . . ?
10-20 Location
10-21 Call (at or number) by phone
10-22 Disregard (cancel)
10-23 Arrived at scene
10-24 Assignment completed
10-25 Report to (meet)
10-26 Estimated time of arrival
10-27 License and permit information
10-28 Ownership information
10-29 Records check
10-30 Danger (caution)
10-31 Pick up
10-32 Units needed (specify such as fire units, ambulance units, and police units)
10-33 Help me quick
10-34 Time
10-35 through 10-39 Held for future: Do Not Use

React

REACT (*Radio Emergency Associated Citizens Teams*) keeps a continuous watch on Citizens Band Channel nine, knowing that many motorists have C.B. equipment which can be very useful to them only if they are able to contact help at any time.

REACT is a full-scale volunteer civilian emergency radio service that meets the modern need to communicate. Team members using their own Citizens Two-Way Radios, monitor Official Emergency Channel nine to assist the public.

Its purposes are to develop the use of the Citizens Radio Service as an additional source of communications for emergencies, disasters, and as an emergency aid to the individual.

It establishes twenty-four hour volunteer monitoring of emergency calls, particularly over officially designated emergency channels, from Citizens Radio Service licensees, and reports such calls to appropriate emergency authorities.

REACT promotes highway safety by developing programs for providing information and communications assistance to motorists. Also, it coordinates efforts with and provides communications help to other groups, e.g., Red Cross, Civil Defense and local public authorities, in emergencies and disasters.

It develops and administers public information projects, demonstrating and publicizing the potential benefits and the proper use of Citizens Radio Service to individuals, organizations, industry and government.

REACT charters local teams who carry out programs implementing the purposes of the nationwide organization.

REACT International, Inc. provides the services of a National Director and staff to coordinate the activities of affiliated teams and to promote cooperation among them in achieving their objective of rendering public service in emergency situations.

For further information about REACT, or if you want to form your own REACT team, write to: REACT International, Inc., Suite 1212, 111 East Wacker Drive, Chicago, Illinois 60601. You can obtain the REACT story through a colorful twelve-minute sound motion picture which describes the REACT program. The film is now available on free loan from General Motors Film Library, General Motors Bldg., Detroit, Michigan 48202. The name of the film is called "Where Seconds Count."

REACT is now a nationwide organization of nearly one-thousand volunteer groups totaling approximately forty-thousand volunteers who utilize equipment in the Citizens Radio Service to monitor Emergency Channel nine and provide local two-way radio communications in response to these emergencies.

REACT teams are prepared to provide supplementary communications in any emergency. Effective local two-way radio communications have proved valuable whenever normal telephone communications have been interrupted because of fire, blizzard, earthquake or other disasters.

Adding Additional Stations

You are permitted to have as many mobile units and base stations as authorized by your license. For instance, you may have two in cars, and one at home; one on a boat and one at the office; or one in the camper and six hand-held transceivers.

If your license doesn't authorize enough transceivers, then fill out FCC Form 505 to modify the existing station license to increase the number of authorized transceivers.

If you want to have more than fifteen transceivers, then file the FCC Form 505 and also write a letter explaining why so many are needed.

Base stations count as mobile units and their location can be anywhere. The address on the license is only the mailing address. For example, if you live in New Jersey and work in New York City and have a summer home on Fire Island, you may have a base station at each location under the same license.

Any of the transceivers may be operated by any member of your family who normally resides with you. When being used for business, your employees may operate your transceivers but only in connection with your business activity.

A separate license is required by family members who do not normally live with you to cover their base stations and mobile units. When it is for business purposes, your employees need their own licenses if they use your transceivers other than in connection with your business.

If you are a volunteer fireman and you install a base station at the fire house and/or mobile units on fire trucks, the fire department must get its own license. The fire department may not free load on your license.

You may have two or more base stations at one location, such as your home. One can be used, for instance, for normal communications, while the other is used for monitoring and communicating on channel nine. You will need two antenna systems or a single antenna system plus a C.B. antenna coupler or a coaxial switch for connecting either transceiver to the antenna.

CHAPTER 9

TYPES OF UNITS

Mobile Equipment. The Citizens Band radios that are intended for use as mobile units are generally very small, very light in weight, and usually draw little power from the battery of the car. They use twelve volts, as do all cars made within the last twenty years (the last fifteen years for the Volkswagens). Power is taken directly from the car. The cigarette lighter is often a good place from which to tap the power source of the car.

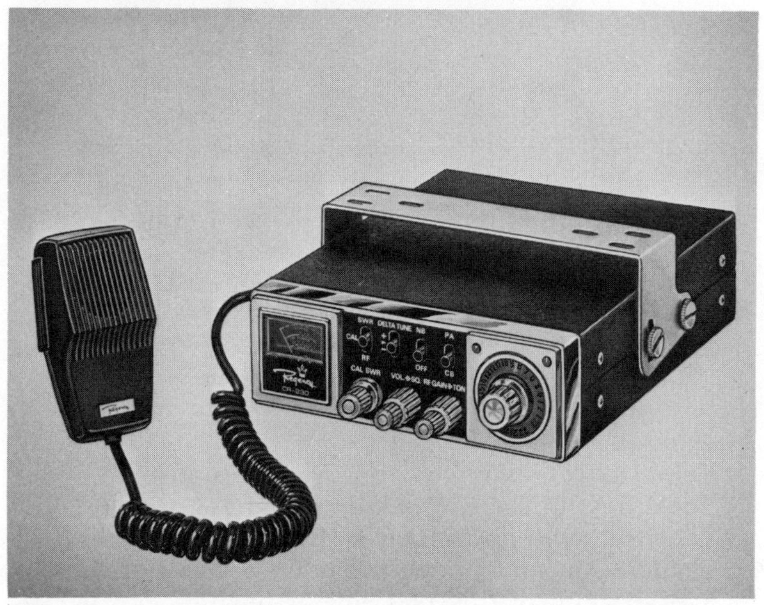

9-1

Antenna. The longer the antenna, the more efficient it is. Remember that a regular "full size" type is about nine feet long. The antenna should be mounted over the center of a flat metal surface such as the roof or trunk.

Short antennas, often called "loaded" antennas are esthetically much more attractive. Every individual has to decide the trade-off between beauty and efficiency for himself. There are twin mobile antennas, popularly called "ears," but these offer more show than added efficiency. Radio amateurs usually do not use twin mobile antennas as a rule.

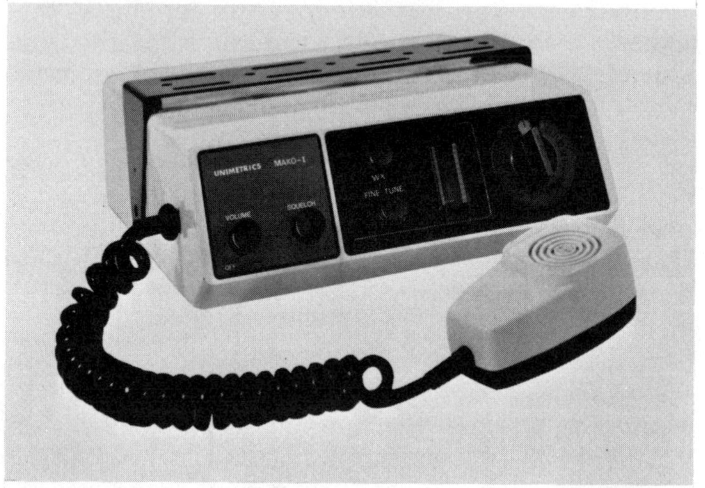

9-2 5-watt, 23-channel transceiver has push-button monitoring of the new NOAA weather stations.

COURTESY UNIMETRICS, INC.

Security. Now, if you are mounting a mobile unit in your car, a special rack can be mounted underneath the dashboard and the unit slid into this. When you are finished driving your car, you can slide it out of this rack and if desired, put it into the trunk until you return to the car, as a protection against burglary.

Security is an important consideration. You know, when you have an antenna sticking out of the back of your car, it is an advertisement to the entire world that you own an automobile that is equipped with a Citizens Band radio. Since you

126

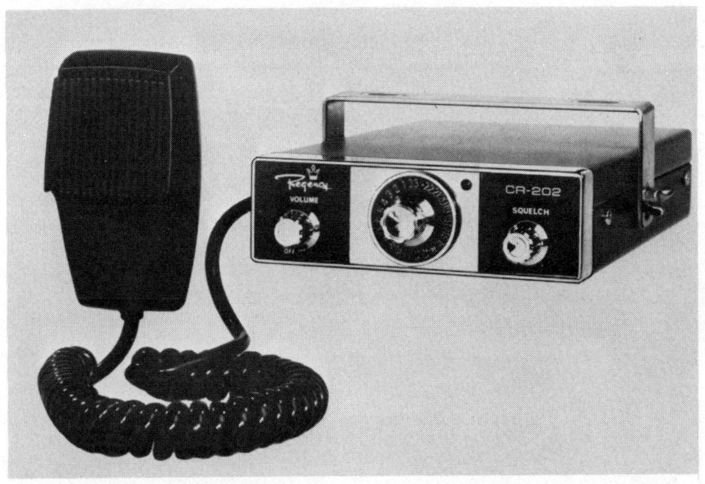

9-3   COURTESY REGENCY ELECTRONICS, INC.

pay three to five hundred dollars for a rig to go into your car, you may as well make provisions to protect such an investment.

If you are planning to leave your unit permanently in the car, you might want to consider a good electronic burglar alarm system for the car doors. Nobody can reach the equipment unless they can open the door first. What a surprise awaits a thief when the alarm goes off wailing through the neighborhood.

Glove Compartment Mounting. In Chapter 5 we spoke of mounting the unit in the glove compartment of the car. It's a good thing to consider since it is out of sight. But you never can tell when that sneaky burglar feels like breaking into your car. No matter how good you locked the doors, he may have the skill to get past locked doors. He just might search through your automobile, and feel around here and there until he came to the glove compartment. Have you felt underneath the glove compartment? Go ahead! Try it! See for yourself what it feels like! And see what it is made of! Surprised? Don't be. Because

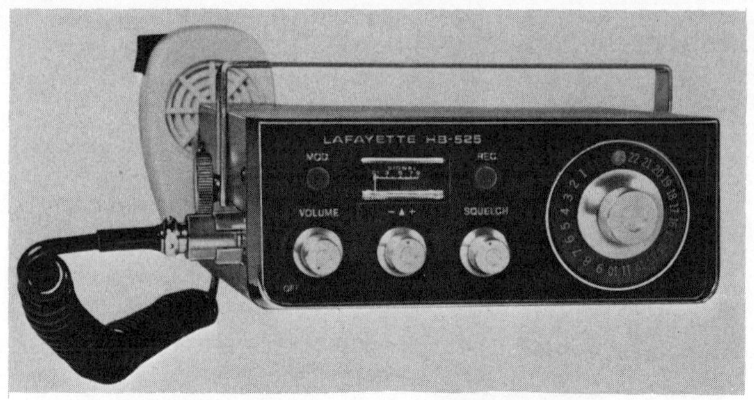

9-4 Mobile transceiver with 23-channels.

COURTESY LAFAYETTE RADIO

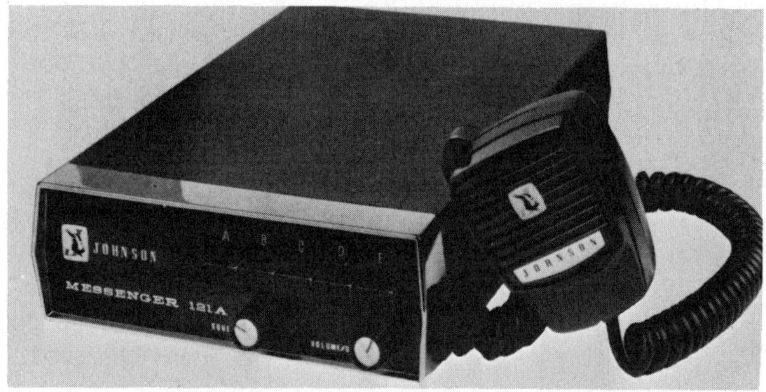

9-5 CB transceiver with built-in automatic speech compression on transmit and automatic noise limiting on receive.

COURTESY E. F. JOHNSON COMPANY

it is what you think it is. Cardboard! And a thief has only to take a razor knife and cut through the cardboard and he's got himself a nice little present.

You can go ahead and report it to the police. Maybe you'll be lucky enough to get it back and then again, maybe you won't be that lucky afterall. Citizens Band radios are negotiable. They are

128

easy to sell. Easy to take out and move. Yes, indeed. These units certainly do represent a big temptation to a prowling thief. So do take the precaution to protect your investment.

Interchanging. Getting back to that rack set-up. If you buy another rack, this second rack permits you to operate the unit in your car or your home. A number of people look upon Citizens Band radio as a hobby, and instead of buying one unit for the house and one for the car, they just purchase a rack, or a base, and they are all prepared to operate in either of the two locations.

For an additional small cost, they could use the same unit in both ways. It's simply a slide-it-out-of-the-car, carry-it-into-the-house affair.

Additional Crystals. The automobile unit is capable of five watts input, and four watts output. Depending upon the price, these units come with a large selection of channel choices. Many units come with a dial on the front and this dial represents an entire range of channel selections, yet there may be only two or three crystals. You then have to purchase the additional channel crystals separately and add them yourself.

More on Installing a Mobile Station. Careful thought should be given to where in the car a unit is to be mounted. Forget about how it will look. There are other more important considerations. One of these is mounting the unit where it will create the least amount of interference to the operation of the car. It might not be the most comfortable place for you to install it, for example, if the unit is going to hit against your knee, or rip your hose or pants, or injure your leg. You'll want to place it off to the side, to prevent any interference with the feet when operating the car. Just make certain your unit is within easy reach, because you do not want to place yourself in an accident-prone position by practically having to lay down on the seat to see what channel you are tuned in to.

There is nothing wrong with installing the unit at an angle. It can be mounted off to one side and still be completely visible to the driver. Mounting it so that the microphone is slung over the steering wheel like a dead carcass is not the best idea, either. Also, avoid placing your unit on top of the dashboard because that is a definite area where it is going to interfere with the visibility of the road. Besides, that type of mounting presents temptation to a would-be burglar. That's like saying, "Look, mister, come see what I've got. You like it? Oh, by all means. Take it. It's a gift for you. Enjoy using it. I'll just buy myself another one. Some day." Nor would it be a good idea to mount the unit directly in front of the passenger side of the car by the front seat. You'll only have to ask your passenger, "Be a good guy, Charlie. Tell me, what channel I am tuned in to?"

Fixed-Station Equipment

Bells and Whistles. Base station or fixed-station rigs are generally larger in size and have more "bells and whistles." These are accessories that may be fun to have, but they do not necessarily enhance efficiency of communication.

Fixed-station rigs have built-in AC power supplies. A solid-state transceiver's natural food is twelve volts DC. Now, while a car battery supplies it very well, the current you get from your house is usually one-hundred and fifteen volts AC.

Some units have *noise blankers,* because some systems, such as an automobile ignition system, emit noise. For example, ignition noise consists of loud but extremely brief impulses that last a few millionths of a second. The noise blanker will shut off the receiver when signal input exceeds a

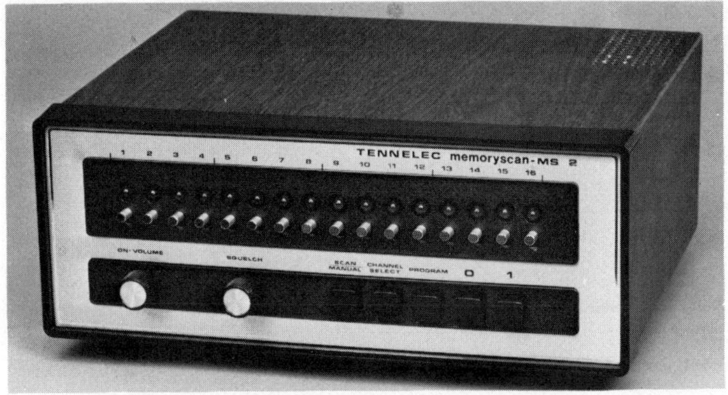

9-6 This scanner lets you change frequencies without buying expensive new crystals and all you have to do is just push a button.

COURTESY TENNELEC, INC.

preset level. Since this type of noise is not there most of the time, the effect is as though the noise just disappears so the desired signal can be heard.

Fixed-station equipment sometimes has a circuit called *range boost,* which in CB talk, is a speech processing unit. When you speak, the speech contains explosive loud parts such as "p" or "b" and quieter parts such as vowels. By limiting the loudness of the loud parts and amplifying the quiet moments these circuits increase intelligibility during adverse conditions such as at the extremes of operable distance. Range boost should be used with care because processed speech sounds just terrible to nearby stations. It all depends upon how much processing you use.

There are many other kinds of "bells and whistles." For instance, some radios are set up for a *"priority channel"* and it usually is channel nine. Regardless of what channel is in use, if a signal comes through on the priority channel, the rig will then automatically switch to it.

I saw one set recently that had a built-in receiver that monitored one of the government stations. It broadcasted nothing but weather reports right up-to-the-minute for airlines, and such, twenty-four

COURTESY E. F. JOHNSON COMPANY

hours a day. You could switch either to that or to the regular Citizens Band.

Still, there is other equipment that includes built-in digital clocks so you can tell the time.

Some of the more ardent C.B. users use amateur radio transceivers as their base rigs. American-made ham transceivers do not have an eleven-meter (27 mhz) position on their bandswitches, because in the United States, Citizens Band radio is entirely separate from ham radio. Remember, C.B. only uses one band of frequencies or channels.

But, in many parts of the world 27 mhz is consid-

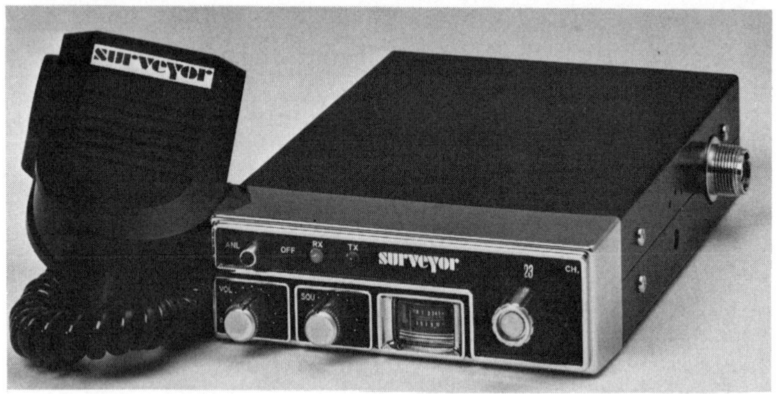

COURTESY SURVEYOR MANUFACTURING CORPORATION

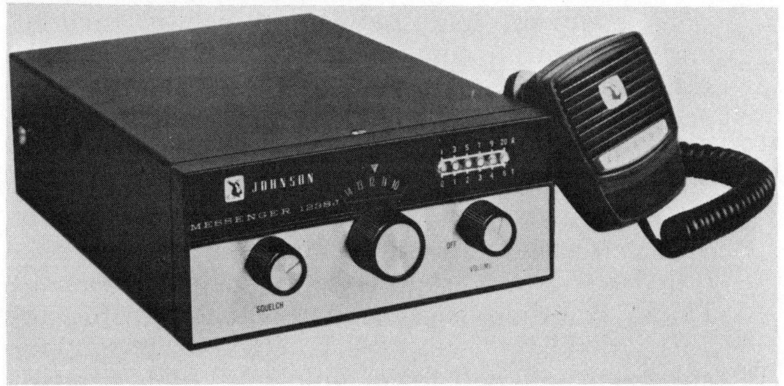

ered to be a ham band, so transceivers made in those parts of the world naturally cover all the amateur bands there, including the 27 mhz. The Japanese make many of these sets.

Also, there are some American electronics manufacturers who market an amateur line without 27 mhz and simply put a different name on the same thing and then sell it as 27 mhz equipment, either with or without the ham bands.

This is done because single-sideband C.B. gear is legally equipped to operate on twelve watts peak power; the amateur radio transceivers put out ten to fifty times that much. Hams are allowed one-thousand watts. This is absolutely illegal for C.B. sets.

Citizen Band users may run either AM or SSB (single sideband). AM stands for amplitude modulation and this means that the information is carried by varying the signal strength of a basic "carrier" frequency generated by the transmitter.

The result of this process is creation of "sidebands" which are the actual information-carrying frequencies very close to the carrier frequency. Ordinary broadcast radio is AM. Notice, when listening to an AM broadcast station that, even when no one is talking, the transmitted signal is "quieting" the receiver.

To compare, tune between broadcast stations and listen to the unquieted noise. SSB, or single sideband, does not have a carrier at all. By generating a "carrier" frequency in the receiver, the full transmitter power is available to transmit information, making it approximately eight times as efficient.

SSB is more difficult to tune. A slightly mistuned SSB signal sounds very much like Donald Duck. But there is also the plus that two SSB signals will fit on a single C.B. channel, one upper sideband and one lower sideband, and that two SSB signals will not *heterodyne*, to produce squealing sounds such as those produced when two AM carriers are on together.

There are some C.B. SSB rigs that advertise sixty-nine channels—twenty-three channels AM, plus forty-six channels SSB, but this is really misleading. That is like three sweaters and three pairs of slacks advertised as nine outfits.

Here is an easy way to explain and understand the single sideband: Imagine yourself as the carrier of two loaded grocery carts, each of which represents a sideband. Then, try to push them down the aisle of the supermarket.

First, you will need a wide aisle, and second, you will have to divide your strength between the two shopping carts. Now, remove one cart, step to the side of the aisle and get behind the remaining cart. You now need only a third of the space you needed before. In real radio you only need about half the space you needed before, since a carrier is infinitely narrow.

There are several schools of thought on buying fixed-station equipment for your home. One is to buy the best that can be afforded. In this way, the first expense would be the last. Others feel they should buy the least expensive unit, to have it do what you can afford it to do and listen in on the band and ask others what kind of equipment they

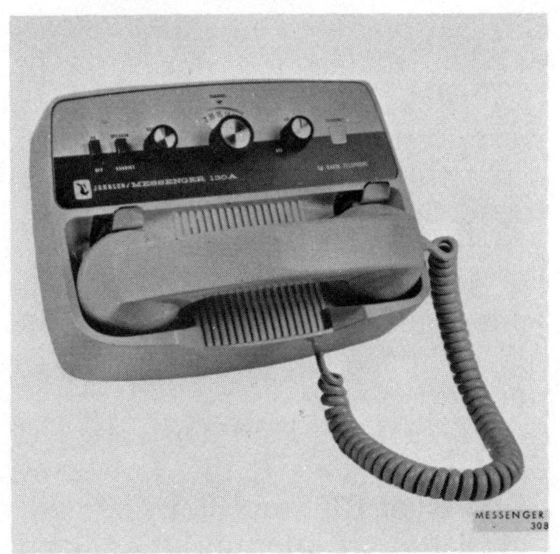

COURTESY E. F. JOHNSON COMPANY

are using. So when you hear a particularly good signal, you can ask the operator of that rig what kind of equipment he has and then you can make a note of it. But don't forget, you can have the best transmitter in the world but if you haven't got a good receiver, you won't be able to hear.

But let's face it. All equipment is not the same. There are better receivers on the market. In fact, there is better everything. It all depends upon how much you want to spend and what you want to use it for. It is a very personal decision and is entirely up to the individual. There is something for everyone. If you have a little extra money to spend, it would be wiser to spend it on another RF amplifier stage in your receiver, in order that you may get better sensitivity, rather then invest money on wood grain panels or chromium plated dials. Such items are just window trimming and have nothing to do with how good your rig can function.

Okay! So the unit in your home is going to remain where it is. In your home! It is not going to do triple duty or double duty as a portable or mobile. It operates on a 110-volt source. You do not need an adapter for your car or an adapter for a battery pack. It will be a plug-in unit. There is no waste if it is just going to operate off the 100 volts.

On the other hand, if you want one unit to do triple duty, then of course it is a lot cheaper to purchase one that is equipped with all three inputs for AC, car, or battery. When using it strictly in your home, it is not wise to waste money on fancy AC-DC power supply systems if you don't actually need it.

How a C.B. Unit Operates. What goes into a C.B. unit? This is the way it works: The *receive mode* is an RF amplifier that takes the weak signal out of the air and amplifies it so we can do other things with it.

The next stage is called the converter. An oscillator, which is tuned by a capacitor, is connected to the variable capacitor which tunes the converter stage, or RF stage. The signal from the RF amplifier and the oscillator are fed to the converter, also called the mixer, giving you an output no matter where those two are tuned, and you will always get a 10.7 mhz output.

As you increase the frequency at the RF stage, you will also increase the frequency of the oscillator so that the remainder, subtracting the one from the other, leaves 10.7 mhz. So, no matter where you are tuning, the output is always 10.7 mhz, no matter what is the frequency of the input signal.

This is fed again into amplifier stages called intermediate frequency amplifiers or IF amplifiers. There are generally two stagews in IF amplification. This is then fed into a second converter which is supplied by another oscillator to give you an intermediate frequency or an output of 455 kilohertz. As you can see, we reduced it from ten points mhz. to 455 kilohertz, and again we have two stages of amplification before this is fed to the detector stage which eliminates all but the audio frequency.

This goes to the volume control which then feeds other stages of audio frequency amplification. Then it goes to the loudspeaker.

Now, in the *transmit mode,* we start with the microphone generating a small current in the microphone, which is fed to an audio frequency amplifier and at the same time the crystal oscillator generates a frequency, which is then amplified by the first RF amplifier stage. Then it goes through what is called a process of conversion or frequency multiplication to bring it up to a frequency of about thirty mghz. for the C.B. unit. This is applied to the final radio frequency amplifier stage at a power level of five watts where it is mixed with the amplified signal from the microphone or modulated. Next, it is fed to the RF amplifier, the output amplifier, and then it goes to the antenna.

Okay! So we've got this whole unit working. Fixed-station equipment should be kept wherever you think you are going to spend the most time with it. Where will you locate it then? Some people like to keep their unit in the bedroom. Some keep it in the den. Still, others prefer to keep the unit in the kitchen. Keep in mind the fact that while it is a small and simple matter to pull out the plug and relocate the unit someplace else in the house, it is not that simple to change the location of the antenna.

Today, most of the units will have on them something that looks like a handle. This gadget can play a double or even a triple role. This device can, indeed, be used as a handle to carry your unit from place to place. But when you are operating it, the handle acts as a stand when placed under the set. This makes it easier to see the dial. It tilts the unit upward at the front. There are thumb screws to lock it into position.

If you are mounting your unit underneath a shelf, you can attach the handle first, then re-attach the unit to the handle, so that the handle can be used as an outgoing bracket to hold it in place. So you can see that the handle is more than a mere convenience for transporting.

You will notice that at the top, bottom, and sides of the case there are holes. When the unit is lying on the table it is a good idea to use that handle as a prop to raise it so as to allow air to circulate through the holes and underneath the unit.

Do not place anything on top of the unit such as writing pads, because this seals off the air holes and prevents much needed cooling ventilation.

Now, there are times, for economical reasons, you just have no choice but to accept a compromise when purchasing a Citizens Band radio. When you know you want a unit to use strictly in your automobile, do not try to get it to do double or triple duty. When you do this you only end up sacrificing one thing for another. And, you will also be paying for it by getting a larger size, and added components that you don't really have use for. With all this, you only pay a higher price. It may appear that you are getting something very efficient, but if you think you plan to use a unit in more than one way and do not, you have only shelled out extra money for nothing.

Radio waves transmitted on Citizens Band radio follow what is called the *ground wave effect.* This means that the signal from the antenna moves along at earth level and is generally limited to line of sight transmission.

Sometimes, particularly in the summertime, when a warm morning is followed by a cool evening, the ionosphere rises creating what is called a *skip effect.* The signals go straight up, bounce off the ionosphere and return to earth from several hundred to thousands of miles distant.

When this occurs, you might experience interference from a station hundreds of miles away. This happens not through any flaw in the equipment, but by mere accident. It is amazing but sometimes you find you are covering a range of thousands of miles with your little five watt power unit. This is called working distance or DX. It can also happen at times when sunspot activity is very high. Solar flare-ups are cyclical, and thus this type of interference is also cyclical.

Generally, when this does occur, two operators exchange QSO cards as confirmation of contact. Thus operators have a written record of the fact that the contact was made.

As wonderful as it all sounds, this is not in keeping with the ways of Citizens Band radio, because you are not supposed to be communicating with another operator who is half way across the world. It is more likely to happen, however, with the rig you use in your home and not the unit in the automobile. That is because you have a better antenna on your home. It is higher up for good reception.

Portable Equipment. When discussing portable units, one is most likely to think about the hand-held ones—the ones with which you hold the loudspeaker to your ear to listen and to your mouth to talk into. Such a unit is a transceiver in which the loudspeaker serves as a microphone when transmitting and as a loudspeaker when in receive.

The loudspeaker consists of a fixed magnet. A cardboard or paper tube fits loosely over the magnet and is attached to the loudspeaker cone which is a diaphragm. Wound around the magnet (which is inside a metal "sleeve") is a coil of wire, called a feed coil. When you put a current through a wire coil, it then magnetizes. Now, there is a fixed magnet inside the sleeve. Because the current going into the loudspeaker varies with the voice modula-

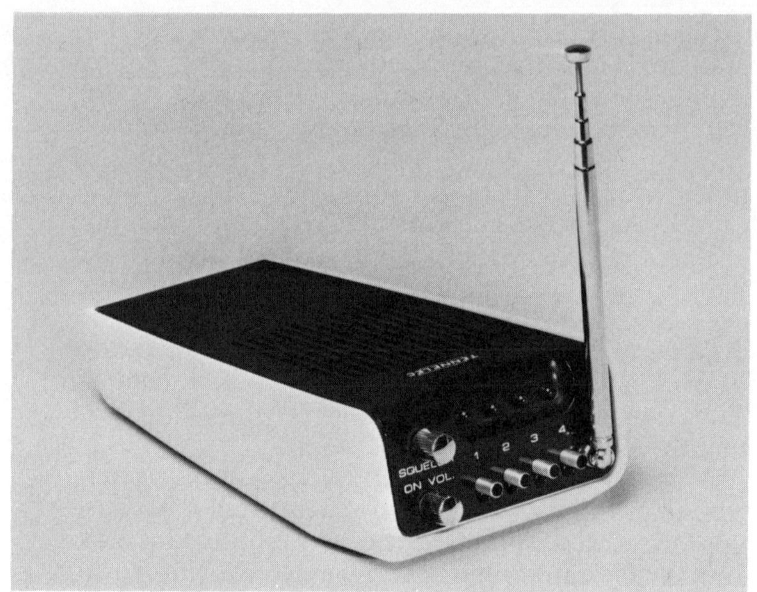

COURTESY TENNELEC, INC.

tion and amplitude, it is going to slide back and forth over the magnet as it becomes more and less magnetized.

As it moves back and forth, it is attached to the loudspeaker cone so the cone vibrates at that frequency and generates all the audio sound you hear. But, if you do not have a current going through the coil, but you speak directly at the cone moving it back and forth with the force of your voice, you are then using your voice to move the cone which moves the sleeve, thus, causing the coil of wire to move back and forth in the magnetic field of the fixed magnet inside the coil and generating current which will vary with the amplitude and modulation of your voice. By taking that small current generated in the coil and amplifying it, you can very well use a loudspeaker as a microphone.

But we are talking about a dynamic microphone in this case, simply because there is no such thing as a crystal loudspeaker.

140

COURTESY SURVEYOR MANUFACTURING CORPORATION

The selling appeal of these units is due largely to its impressive appearance. When you shop you will find some with what looks like expensive chromium trimming. But if you take a knife and scrape through the finish, you will find *flashed metal*. Flashing is a process of applying a thin layer of metal to plastic. It is very much like paint, maybe a little better—but not much. So the end result is the appearance of metal when it is really not.

Another factor that sells portable units is their small size. People want something they can stick in the pocket of a shirt or whatever. However, it is not always that way.

When it comes to engineering, you cannot get something for nothing. If you want a portable unit, one with high efficiency, and which has a good receiver section with five watts input, you have to make room for all the parts that must be crammed into such a small cabinet. So, some of the portables do tend to get just a wee bit bulky.

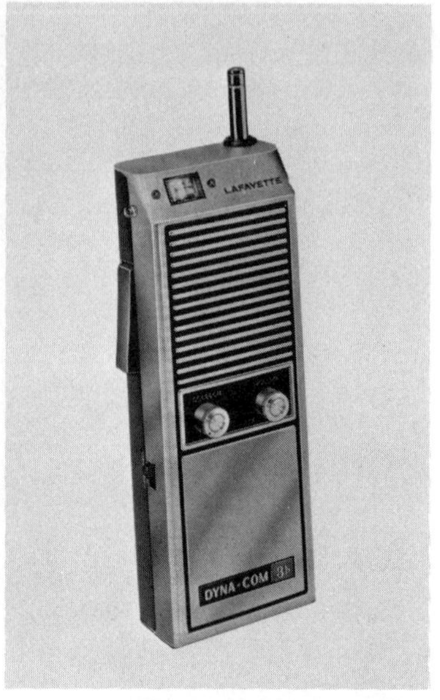

COURTESY E. F. JOHNSON COMPANY   COURTESY LAFAYETTE RADIO
                                        ELECTRONICS CORP.

One problem with a portable unit is the weighty batteries needed when dealing with a high-power range. However, this type of unit comes in various power input ranges and with many kinds of features, some of which you would not expect to find with it.

You already know about the one small children use—the type that operates below one-hundred miliwatts. But there are other portables that come in the three-watt input range and still others with full five-watt power. It's like with anything else. You get what you pay for. Some portables require a change of crystals if you want to change channels. And some portables are rechargeable.

Lafayette has a portable unit which can be recharged by simply plugging it into a special base they have, which plugs into the wall. By plugging this device into your home, you can use this portable unit as a fixed station. It also comes with an

adapter which enables you to plug it into the cigarette lighter of the automobile. With these variations on this kind of unit, the portable need not be
something you only carry with you.

The convenience of the portable unit is that it is
completely self-contained. Many of them use electronic condenser microphones so that you do not
even require an external microphone. When you
are selecting one of these units, find one that is
light in weight. Some portables look about the
shape and size of a fixed-station unit, but have
handles. But take care that the one you are choos-
ing is not too bulky or heavy.

It should feature a rechargeable battery, offer
operation on internal batteries, external portable
batteries for use in a car or an external source of
one-hundred and ten volts, sixty hertz AC. Consider its size. If you are looking for one of those
really small pocket units, it will be difficult to find
one that comes with a five-watt power input. It's
hard to put all that power into such a small package.

COURTESY E. F. JOHNSON COMPANY

So what is the advantage of owning a portable
unit? Well, for one thing it will operate where you
do not have a twelve-volt supply source handy and
there are places where it would be just the thing
you would need. Perhaps you could take it along
on a field trip or a hike, or perhaps camping out-of-
doors.

Batteries. When it comes to batteries, or anything else for that matter, the optimum design is
one that is most economical. There is the voltaic
battery which is the kind you have in your car.
This kind has lead plates suspended in a solution
of sulphuric acid and held in a container of heavy
plastic.

This battery has to receive a charge in order for
it to function. Now when you operate your car, the
generator works through the voltage regulator to
apply voltage to the battery to keep it in a fairly
charged condition. It usually consists of a series of
cells, each one providing one and a half volts and
two "red" plates per cell.

When these batteries are in use lead flakes off
each plate and falls to the bottom of the pool of
sulphuric acid. When enough lead has flaked off,
it bridges the distance between those two plates in
the battery, shorting out the circuit and the cell
bars break.

To prevent this from happening, acid-resistant
fiberglass blankets are placed over the red plates.
These blankets are porous, this preventing the red
plates from flaking overly fast.

When the red acid battery is working it emits
fumes. These fumes give off hydrogen as it is
breaking down the sulphuric acid into hydrogen
and sulphur. It is quite dangerous to have around
an open flame. It can explode.

The LeClanche cell is the next most common
type of battery. This is the ordinary dry cell such
as is used in a flashlight. This type of battery con-

sists of a zinc can with a carbon rod at the center. The balance of the zinc can is filled with potassium dioxide or sal ammoniac (ammonium chloride) in a paste form.

Upon the immediate assemblage of this battery, it is capable of generating one and a half volts of power, the length of time depending upon its size. The number six battery is approximately six inches high and three inches in diameter and has two screws at the top. This cell will generate power for a reasonable amount of time.

The size D cell will last longer than a size C cell and the size C cell will last much longer than the size AA cell. The larger the cell the more current is available. You can connect these batteries in a number of ways.

You can connect them in a series that is plus-to-minus-plus-to-minus-plus-to-minus and the voltage will add. Each cell generates one and a half volts so that if you were to connect two of them in a series, you would have a total of three volts. Current availability remains the same.

If you connect them in what is called a parallel arrangement, all the pluses and minuses are connected together. At the terminal the battery will have one and a half volts, but you will have added the additional current capacity of each battery so that the total will last longer.

If you take a nine-volt transistor battery apart, you will find it is nothing but a series of six small cells which are connected in a series and giving you a total of nine volts. Each one is generating one and a half volts.

The problem with these LeClanche cells is their limited shelf life. It would be a good idea to go to a reputable store where they keep a large stock and have a big turnover, so that you can be reasonably assured of getting a fresh battery. This type of battery starts to go bad the minute it is put together at the factory. For all you know, you could buy one that was sitting either in the warehouse or on top of the dealer's shelf for a long time. Months! There is every possibility that the new battery you just purchased is dead.

An improvement over the LeClanche cell is the alkaline battery. Now this kind of battery has an unlimited shelf life in that it will not begin to go bad until it is put into service. The alkaline battery can sit on the shelf for ten years and nothing will happen to it. Not until you connect it into the circuit. Then, and only then, will it begin to be used. So this kind of battery will last ten times as long as the LeClanche cell. Although these are

more expensive than others, they are a much better value for your money.

Then there is the nickel-cadmium battery, in which the outer container is made of nickel and the inner core is made of terminal cadmium. One advantage of this cell is that it is rechargeable. You can use it, put it into a proper recharging circle, and then bring it right back up to par.

Still another type of rechargeable cell is the *nife cell* which is made of nickel iron. This cell works in essentially the same way as the nickel cadmium does. The mercury cell, is still another battery we ought to know about. The mercury cell is used a great deal in photography, electronic wrist watches, and in hearing aids. They can be made very very tiny. The characteristic of this cell is that it is either good or bad. It does not decay slowly. It works at full pitch until it stops working all at once. That is why it is so useful in photography. In photography, light meters are used and because the mercury cell is used, remembering that this kind of cell remains steady, it keeps the meter readings accurate. No false readings due to a failing battery. If the battery goes bad, it does it right there and then. It just stops working. These have been used to some extent in the meters in Citizens Band radio.

CHAPTER 10

THE FUTURE OF C.B.

Since the license fee for Citizens Band radio was dropped from twenty dollars to just four, approximately one-hundred thousand more applicants a month have literally been pouring into the Federal Communications Commission.

What then happened, was that amateur licenses were getting months behind. Commercial license operator renewals kept getting lost in the shuffle and finally, C.B. licenses began to take so long that many stations began using their handles or nicknames instead of their call letters.

And so, the Class-D Citizens band began running out of control. Hobbyists took over C.B. Manufacturers and amateurs fought so hard against reconstructing Class-D, that the Federal Communications Commission gave up.

Sometime back in February, 1975, the FCC wanted to reconstruct the Class-D band. They decided to open up single-sideband operations only, except for channels nine and eleven. They wanted to control the antennas, too. The FCC began to really hit hard by giving out heavy fines to people who used their units for hobby purposes.

But at the same time, the FCC also spoke of changing amateur radio licenses. Part of this proposal was for what was called a communicator license which would allow the use of unskilled communicators. There would be no code-speed plus a much simplified written examination.

Ham radio operators complained bitterly to their Congressional representatives. All those complaints brought about a postponement of the restructuring.

So, with the hobby restrictions off, C.B. users and C.B. manufacturers were elated. C.B. enjoyed a boom. Now more than ever everyone was trying to get into the act—department stores, automobile service centers, auto supply shops, discount houses, tape player specialty shops, etc. What is badly needed at the present time, is some kind of expansion of Citizens Radio Service.

It seems as if the future of C.B. lies in the Class-E Citizens band. You see, Class-D is for the hobbyists. Class-E would not only be used for business and personal use, but would also act as a buffer against the hobbysits. In that way, a true Citizens Band Service would be provided.

The license would most likely be given by the dealer when a unit was bought. He would then file the papers with the Federal Communications Commission and he would probably put the cost of the license in together with the purchase price of the transceiver.

Because the identifier is coded to the license, violations would not be difficult to detect. Therefore, a transmitter can communicate only with another station with the exact identifying code.

The fact that all Class-E transmitters would be licensed at the time they are bought, leaves the hobbyist only one way of getting one—he can steal the equipment, which is, of course, a crime.

A Class-E unit, of perhaps ten to twenty-five watts in output, for instance, and having about ten channels with which to choose from, would cost below two-hundred dollars. It would probably cost approximately fifteen dollars more for the antenna equipment. So the price would be almost as same as for the Class-D system.

However, it seems as though Class-D would still remain, for there does not seem to be any legal way to eliminate or make any changes in what has become a legal unlicensed amateur band.

Maybe one day in the future, cars might come already equipped with a transceiver at the time of purchase. Therefore, the channel nine emergency service would be provided automatically!

150

In recent years there has been an increase in the availability of low-priced C.B. transceivers. While not that terrific in sensitivity or selectivity, those units became very popular because it cost just about one-hundred dollars, a price especially right for students and young children.

Today, one notes in practically all of the major newspapers, from a half-page, to as much as a full page ads for the promoting of C.B. radios. Prices are beginning to go from two-hundred dollars to five-hundred dollars because C.B. hobbyists are always on the look-out for more and more improvements.

As for the Class-D restructuring with its additional channels, there probably won't be too much future, simply because there is only a single-sideband, and the cost would be too high. Small businesses would not be able to stock them and the personal users would not really be able to afford them.

Manufacturers, dealers, and users feel that extra AM channels should be added and reserved for the business and personal user, which is what the Federal Communications Commission wanted in the first place. In this instance, contacts would be made only between stations holding the same license.

The FCC would insure absolute freedom from C.B. hobbyists with built-in identifiers. Then the hobbyists would slide up the band to the new channels. So if the new channels were reserved for business and personal use, Class-E would not really be needed. And so, the future of C.B. will probably be determined by new frequencies and how they will be used, and not by how they are initially set up.

With the way Citizens Band radio is growing in popularity, the Federal Communications Commission will grant approximately three million new licenses in 1975. There are a number of reasons for its wide-spread growth and its appeal. One of the reasons is the use of Citizens Band radio by some law enforcement agencies.

Some other reasons include the reduction of license fees to the current four dollars; the energy crises; and the highly publicized use of Citizens Band radio by the truckers.

Today, the usefulness of Citizens Band radio is simply overwhelming. A man or a woman driving along on a lonely road need not actually be alone anymore. Not as long as their vehicle is equipped with a rig.

Armed with C.B. transceivers, many people have banded together to keep their neighborhoods safe. They have managed to reduce telephone calls to the police by eighty percent. In this way, older persons feel more comfortable on the streets at night. For those citizens who are confined to the house or who prefer to observe from their windows, Citizens Band radio also comes in handy.

We begin to note more and more news stories relating how Citizens Band radio came to someone's rescue. A man in a shopping mall was about to get out of his automobile when he was accosted by two assailants. One of them threatened him with a screwdriver. The man managed to knock the weapon from his attacker, and he had just enough time to call for help on his Citizens Band. Only seconds later, the police arrived on the scene and apprehended the criminals. So C.B. was actually responsible for having saved that man's life.

In this regard we ought to consider the growing overcrowdedness of the channels. It would be a frightening thing if the day should ever come when this fast-growing situation prevents a person who needs help from getting through to those who can help. Citizens Band radio is here to stay! It really is! It is up to those of us who use the channels, to get in there and actively do something. And in unison!

American Radio Relay League. This is a group of amateur operators who have appointed certain stations to monitor the airwaves and report anyone who is breaking the rules of C.B. When the league is apprised of this, they send out a warning. They do this on the first report.

If they have to send out a second warning, it is a much more severe one. Should there be a third warning, the person who keeps on breaking the rules will be reported to the Federal Communications Commission. The report will consist of that person's identification, the description of his violation and his location. That person will also receive a copy of the report that is being submitted to the FCC.

So you see, what this whole thing really boils down to is this: if Citizens Band users want to stay on the air, they must then help themselves by helping the Federal Communications Commission. How do they go about doing this? By policing themselves. People who get on the air and who think they can take all the time in the world, and take it selfishly must be curtailed. They must be made to realize that there is enough room for everybody, if they just learn to be courteous and polite. While the Radio Relay League doesn't actually have the sanction of the FCC for their work, still it has been proven to be very helpful.

Repeater Stations. The repeater station is worthwhile considering. Two-meter amateur radio is limited in range, the same as Citizens Band radio, and this is due to its high frequency and line-of-sight transmission characteristics.

Ham operators all over the country got together and established something called "repeaters" which operate at maximum power and at specified frequencies. How is how this works.

Suppose you are a ham operator living in one town and you want to try and reach a fellow ham operator who is living in another town. Also suppose that because of some kind of interference you cannot reach him directly. What you would both do is tune to the repeater. Your signal is received by the repeater, re-transmitted and then received by the other ham.

C.B. users could set up such repeater stations in all the metropolitan cities on specified channels. They could, therefore, extend the usuable range of their transmitters, too.

There could be another means of softening the problems of overcrowded airways: reduce the amount of power output the rigs are allowed to use.

Perhaps a coaxial resistive element could be put in line with the antenna to reduce the output power to two or three watts. Although you would not be able to operate too far out of your backyard you would have less interference from those who are using the five watts. You do not want to talk to them anyway. However, this probably is not the best of solutions because the Federal Communications Commission would have no way of checking. So, what good would it do to make rules that cannot be enforced?

Radio Relay. This is another project that amateur radio operators have set up which seems to be working out quite well. Perhaps C.B. people could think about using it. Suppose you want to relay a message to a friend or even a relative. But you live in New York and you want to communicate with California. There is what is called a "network" which meets two nights a week and this is done right over the air.

You would go about sending your message by finding the furthest west member who would then take and relay that message for you. That station would pass it on to another station, and by repeating this process many times, your message would arrive, by the end of the evening, to the town nearest to your destination. From there, the operator would make a collect call to the receiving party and relay your message on the telephone.

Citizens Band radio is changing the social patterns of the American people. At one time people had to be face-to-face in order to communicate with one another. Those with whom they communicated were usually people they knew very well. At that time, mobility was low, and citizens often lived in one town for their entire lives.

When Alexander Graham Bell came along with his wonderful new invention, the telephone, everyone spoke to whomever they pleased, and despite their new broad scope, they still communicated often with those they new quite well.

Now, people are learning they can communicate from their cars, too, and not just from a fixed location in a building. People can also communicate from boats, sidewalks, open fields, and from just about anywhere. Instantly!

This facility of instant communication from anywhere to anywhere, from anyone to anyone, is preparing us for new technology just around the corner. Already supermarkets are beginning to use automated checkout facilities where a computer device reads prices, rings up the cash register, calculates the change, keeps track of inventory. Right now banks are working hard on what is called the "checkless society" in which the employer will send the bank the salary records for its employees on a computer-readable magnetic tape. The electric company and the supermarket plus everyone else will electronically order the bank to deduct the right amounts from your bank account and then transfer them to their accounts.

Citizens Band radio? It has become a part of the lives of the American people and has literally interwoven itself into their everyday lives.

GLOSSARY

AGC—Automatic Gain Control—A system to eliminate noise interference.

Alkaline battery—Battery with unlimited shelf life.

ANL—Automatic Noise Limiter—Suppresses external noise.

Base station—A transceiver in a fixed location usually used to communicate with mobile units.

Beam antenna—A directional antenna that concentrates most of its energy in one direction. In receiving, it is much more sensitive in one direction than other types of antennas.

Breaker—C.B. operator who breaks in on another operator's conversation.

Call sign—Station identification assigned by the FCC with a license.

Carrier power—RF power output of an AM transmitter that is not being modulated.

Clarifier—Controls the best clarity of a received signal.

Coaxial cable—The transmission line for interconnecting a transceiver and its antenna, consisting of two conductors with a common axis.

Crystal—A plug-in package having a precisely cut, paper-thin slab of quartz crystal which vibrates at one specific frequency. A transceiver needs twice as many crystals as its amount of switchable channels.

Delta Tune Control—Fine tunes the received signal.

Diode—Solid-state device used as a rectifier.

ERP (Effective Radiated Power)—The measured effective power which is determined by the transmitting power, antenna type and antenna location.

FT (Fail-safe Transistor Protection)—Automatically protects the transmitter in case of a broken, shorted, or disconnected antenna.

Handle—Nickname used by a C.B. operator.

L/DC (Local/Distant Control)—For receiver range adjustment.

Miliwatt—1/10 of a watt.

Mobile station—A unit temporarily installed at a fixed point.

Modulation—A change in the frequency, amplitude, or phase of a signal in order to superimpose intelligence on that signal.

Negative ground—When the negative battery terminal of a vehicle is grounded to the body and the frame of a car.

Noise Blanker Switch—For suppressing noises.

Omnidirectional antenna—An antenna which radiates equally in every direction.

Pack set—A transceiver you can carry by its handle.

P.E.P. (Peak Envelope Power)—The power which is generated by a single sideband transmitted when it is modulated.

Piezoelectric—crystal of Quartz or Galena which is used to control oscillator frequency.

PTT—Push-to-talk—Most microphones are equipped with this function also known as press-to-talk. A button is pushed, turning the transmitter on. When it is released, a signal can be received.

QSL (Confirmation card)—A post card C.B. operators send to each other when they have talked together over the radio. (Must be requested)

REACT (Radio Emergency Associated Citizens Teams)—A volunteer group of citizen radio operators who serve their communities with emergency and highway safety communications.

RF (Radio frequency) **gain control**—For receiver sensitivity adjustment.

Rig—Slang talk meaning transceiver or truck (double meaning).

Scanner—A specialized receiver that instantly tunes in on a channel being used and locks in on it, and then scans again at the end of a transmission.

Skip—A term used when radio signals can be broadcast over a great distance when the signals are reflected back to earth by the ionosphere.

S Meter—Monitors incoming signal strength.

Squelch control—For quiet standby.

S/RF Meter—Monitors incoming signal strength and relative power output.

SSB (Single Sideband)—A method whereby only half of an AM signal is utilized, thereby providing twice the given space on any channel.

SWR Meter (Standing Wave Ratio Meter)—Relays antenna system's performance.

Talk power—Amount of effective modulation.

10-Code—Abbreviated radio communications talk.

Transceiver—A transmitter and receiver built into one cabinet and using many components in both functions.

APPENDIX A
What's on the Market Today?——Accessories

E. F. JOHNSON COMPANY

C.B. matchbox adjusts antenna line for lowest VSWR, peak efficiency. Handy for adjusting antennas. $21.00

Accessories for hand-held portable. Battery charger plugs into Messenger 109 for charging without removal of battery module. $37.50

In-Converter permits use of any negative ground Messenger mobile radio with a positive ground system. Can also be used to operate radio from 6 volt systems. $29.50

External speaker can add increased clarity under noisy conditions. $16.95

Transceiver checks C.B. transceiver performance 10 ways. Reads power output in watts and modulation in percentage. Has built in RF and audio generators, crystal activity checker, SWR meter. Also features built-in antenna for field strength meter readings. $65.00

GC ELECTRONICS

Magnetic mike holder clips easily to all standard mikes without drilling holes. Magnetically attaches mike to transceiver case, auto instrument panel, any steel surface eliminates fumbling.

"Lucky" antenna spray. 3 oz. aerosol. Weathering deteriorates any antenna system from the moment it is installed. This spray can be used on connectors to "seal" them from water, air pollution and salts. Maintains constant SWR for years. Dries clear. Use on any mobile system or base.

Compact bumper mount. Bumper mount for compact and subcompact cars. Designed for smaller bumpers, such as those on Vegas, Pintos,

VW's, and many other imports. Accommodates whips with standard ⅜"-24 base threads. Stainless steel strap holds nickel plated mount securely.

C.B. Co-Phaser. For mobile installations-no special harnesses, no special cut lengths of wire-connect any two C.B. antennas to your transceiver for better gain. Add another antenna to your present system and move-up to a Co-Phased System. Run coax cable where you want it and cut it to any length you desire. No extra coils of wire to hide. Fine tune control for perfect SWR balance. More consistent communication without "Dead Spots" or fading when changing direction of travel.

Scanner cable. Shielded "Jumper" for connecting police monitors, FM converters-numerous units. Heavy shielded cable with Motorola plug each end. 24" length.

Weather resistant RG/U coax cable. New, low-loss type PL-259 connector one end-spade lugs on the other. For mobile installations using trunk mount, mirror or bumper mount.

C.B. omni-meter. An excellent in-the-field multi-purpose meter for antenna installation, tuning or repairs. Compact and rugged, can be used while installing mobile or base station antennas and peaking transceivers for best performance. Measures two power ranges. VSWR in two ranges and relative-field strength (separate antenna included). Rear mounted connectors allow use as permanent base monitor.

GOLD LINE

Coaxial switches. A Gold Line Coaxial Switch attaches easily, and can be mounted anywhere. With a flip of the switch you can change from one antenna to another.

Inline watt meters. Reads true output power in watts. Stays in line for constant monitoring. Negligible insertion loss. Easy to use and very accurate.

SWR Bridges. Handles 250 watts average power. Units are supplied with a universal mounting bracket for inline applications and are designed for 50 ohm impedance lines.

Grounding straps. Fights noise. Static discharges interfere with good radio communications. Eliminate static noise caused by poor ground between engine and frame, body and frame and many other areas where good metal to metal connection is necessary for proper grounding.

Lightning arrestor. Protects valuable equipment against lightning. Eliminates static build-up. Employs positive gap principle. No screws to adjust. Silver plated. Compact.

LAFAYETTE RADIO ELECTRONICS CORPORATION

Lafayette antenna impedance meter system. Provides the most precise and accurate impedance matching available for optimum performance. Helps you get the most efficient SWR and power output by precisely matching your antenna to your rig. 9-volt battery. $69.95

Antenna coax slide selector switch. Easy to operate 2-position antenna coas switch. 50 ohms impedance. Nickel plated UHF connectors with patented slide type transmission line. 2 mounting brackets. $12.95

Broad-band balun. Model matches 50- or 75-ohm unbalanced to 50- or 75-ohm balanced load. $12.95

Dummy antenna load. Rated 100 watts continuous, 1250 watts with fan cooling. 52-ohm non-inductive load for easy RF power. $22.95

LEADER INSTRUMENTS CORPORATION

LAC-895 antenna coupler. Built-in SWR and in-line powermeter for accurate measurements in 5 bands from 3.5 to 28 mhz. SWR measuring circuit facilitates matching adjustments. Incorporates 2 switchable direct output circuits. Helps obtain cleanest receive signal and transmit maximum power output.

LDM-815 Transistorized Dip Meter. Checks receivers, transmitters, antennas and other circuitry. Useful as a high sensitivity absorption frequency meter.

LIM-870A Antenna Impedance Meter. Adjusts antenna over an extended period of time to achieve optimum matching. The instruments measure feeder impedances. Reads receiver input, cable and filter measurements.

LA-31 3" scope adapter. For output monitoring is designed for use with Leader's LBO-310A or any other scope with direct deflection plate connections. You can monitor waveforms and output power of single sideband and AM transmissions by using the LA-31.

MOSLEY ELECTRONICS, INC.

Beam stacking kit. ASK-345—$79.95
Dual coax wallplate. C-2PK—$10.35
Mobile ant. magnetic mount. MG-1—$41.70

MURA CORPORATION

MURA C.B. microphone connectors. Priced from 88¢ to $1.85. For more information you can write to them at 50 South Service Road, Jericho, N.Y. 11753

PACE

P5252-Crystal pair. For transmit and receive for C.B. 10-2 and C.B. 100 ASA. (Channels 1-23)

Carrying cases.

Speakers. P5503A-All weather 5″ Trumpet Speaker—7.5 watts.

P5807-charger. 50mA for Nicad batteries. Used in handheld units.

P5809-AC Adapter. 300 mA, 4-way plug

P5832-Portable battery field pack with antenna. Holds 8 "C" size batteries.

P5513-Earphone. For handheld units.

SURVEYOR MANUFACTURING CORPORATION

6″ x 9″, 10 ounce coaxial speakers. 10 watts, RMS, 8 ohms. impedance. Weatherproof. Practically distortion-free sound.

Door Mount, 6″ round air suspension speakers. Model SAD-610. In-door, air suspension speakers. Matched. 10 ounce Surveyor magnets, soft, padded, snap on grills, black cone whizzers, 7 watts.

Model 5020 dual-chain bumper mount. With double lock links for adjustment to any bumper size. Protective vinyl chain cover—⅜″-24 thread. Holds any length whip and spring.

Model 5023 VHF Rubber duck antenna. Ideal for rugged outdoor use to replace wire or telescopic antennas on "pocket scanners"—totally insulated, can't be shorted out.

Antennas

ANTENNA, INCORPORATED

The Equalizer. Model 12510-Mobile base loaded rooftop antenna, ⅜″ snap-in-mount, spring,

34" stainless steel whip. $21.25

The Leveler. Model 12520-Mobile base loaded rooftop antenna, ⅜" snap-in-mount, spring, 33" fiberglass whip. $24.40

The Enforcer. Model 13111-Mobile base loaded gutter mount antenna $28.88

The Big Boomer. Model 14111-Dual base loaded truck mirror mount antenna, heavy duty chrome plated mounting bracket, phasing harness. $49.00

The Howitzer. Model 22530-Base station antenna, ¼ wave, 3 radials, accepts PL-259 connector. $14.25

GC ELECTRONICS

The Grabber. Top loaded whip antenna—48" fibre glass top loaded whip. Highly sensitive. Not easily broken. Heavy, copper-clad whip can be bent and return upright without harm. Loading coil is tunable for best SWR setting.

Interceptor Mach I & II antennas. Base loaded weatherproof design assures excellent 50 ohm match in any kind of weather. Snap-in-mount. Complete with 16 ft. coax, connector, stainless steel radiator and shock spring.

LAFAYETTE

Lafayette 2-meter mobile power gain antenna. 51"-⅝-wavelength 2-meter antenna has 3.4 dB gain. SWR of better than 1.5:1, power rating of 100 watts FM. 17-ft. cable PL-259-trunk mount. $27.95

Lafayette 2-meter gutter clamp mobile antenna. High efficiency antenna with heavy-duty spring clamp for temporary or permanent installation. 19" stainless steel whip. 23" long overall. With cable and PL-259 connector. $12.95

Lafayette 10-80-meter vertical antenna. Turn-

able, irridite-treated to MIL specs. 18-ft., knocks
down to' 5-ft. 1000 watts AM or CW, 2 kW PEP.
Omnidirectional. 3½" air-wound coil. 52 ohms.
$29.95

MOSLEY ELECTRONICS, INC.

Black Widow mobile antenna. BW-0-$21.00
Tarantula truck antenna. CT-1-$24.95
Mosley SWV-7 Fixed vertical SW antenna. Has
six trap assemblies, weatherproof phenolic casing.
$72.95

THE MURA CORPORATION

*The Sky Hook, Model CBA-1-Mura gutter-clamp
antenna.* Attaches to gutter of any car, truck, ambulance or other vehicle without need for screws
or drilling. 9-foot coaxial cable leads from antenna, terminating in a standard PL-259 type plug
connector. Overall only 21-inches high, including
the tunable 12-inch whip. Center-loaded design
for improved performance with mobile C.B. transceivers.
The Range Finder. Model CBA-2-Mura trunk-
lid or roof-top antenna. Attaches to trunk lid without need for screws or drilling. May be used op-
tionally as a roof-top antenna after drilling a hole
for cable access and for through-the-roof screw.
45-inch antenna is tension spring mounted on a
rubber-rimmed chrome base. 3-inch diameter.
15-foot coaxial cable is terminated with a standard
PL-259 type connector. Base loaded design for improved performance with mobile C.B. transceivers.
$24.95

PATHCOM, INC.

P5601-Base antenna. Ground plane base antenna with 3 radials. Complete with 50'RG8 cable and PL-239 connectors. 25–50 mhz range.

P5615-Mobile antenna. Double mobile truck mirror mount antennas-2 fiberglass whips with tuneable tips. Complete with mirror mounts, cable harness, and connectors.

P5650-Portable Base/Mobile antenna. For use with P5833 portable battery field pack or temporary 5 watt CB base station operation.

SAXTON PRODUCTS, INC.

Model 9760. Trunk mount antenna with 48" top coil loaded, tuneable-fiberglass antenna. Deluxe rapid grip mount provides permanent installation in seconds without drilling holes. Installation kit contains a custom designed adapter which allows for removal of antenna when entering car washes or for the prevention of theft. Contains 20' RG-58/U white cable with PL-259 connector. $31.95

9761-Mirror mount, with 48" top coil loaded-tuneable-fiberglass antenna. Contains rugged engineered mirror mount for mounting on square or round tubing. Included in installation kit is adapter which allows for removal of antenna for washing or prevention of theft, also 10½" of RG-58/U white cable and PL-259 connector. $28.95

9762-Twin-trucker antennas. Comes complete with two 48' top coil loaded-tuneable-fiberglass antennas, two rugged mirror mounts with adapters which allow for removal of antennas for washing or prevention of theft. $40.95

Hustler

NEW-TRONICS CORPORATION

Dual 51" mirror mount truck antenna system. Model HTM-1

Deluxe dual rain gutter C.B. antenna-Model DGG

Base station antenna with chimney clamps-Model 27T

Roof mount 48″ base loaded antenna-Model XBL-3

Trunk lip mount 48″ base loaded antenna-Model XBLT-3

Super Swamper-Base station antenna-power gain-Model 27TD

Trunk groove mounted 48″ center loaded antenna-model TGF-271

Microphones

THE ASTATIC CORPORATION

Model D-104-Crystal. $43.50

Model GDN-50-Dynamic. $68.00-Complete with grip-to-talk desk stand.

Model 539. Noise cancelling ceramic-push-to-talk switch, cancellation of background noise for sharp, clear speech.

GC ELECTRONICS

Model 18-010. Power hand mike—Designed especially for C.B. transceivers, mobile or base rigs. Power amplified modulation punches through "skip" and interference to increase range. Solid-state circuitry withstands temperature extremes and prolongs battery life.

Model 18-000. Power base mike—Excellent with AM and SSB. Uses standard 9 volt battery. Battery life: average 200–300 transmitting hours.

Model 18-030. High impedance hand mike—Built to withstand temperature extremes. 250K ohm ceramic element.

MURA

The Economy-DX-114. A quality dynamic microphone. Push-to-talk switch is provided to activate the relay or electronic circuitry in transmitter.

The Loudmouth. DX-116 with gain controlled amplifier.

The Knock-about. DX-115D and DX-115C-ceramic cartridge.

The Big Mouth. DX-117 with two level amplifier. User can set to sensitivity required. Dynamic mike.

The Silencer. DX-119—Noise cancelling dynamic mike where background noises from machinery, office equipment, airplanes, etc.

SHURE BROTHERS, INC.

Shure series No. 200 ceramic. Unaffected by severe temperature and humidity changes. High impedance. Rugged.

Shure series No. 400 controlled magnetic. Balanced armature, rugged, stable, high output.

Shure series No. 500 dynamic. Moving-coil microphone. Superior in frequency response to ceramic, crystal, carbon, and controlled magnetic units. Unidirectional, omnidirectional.

SURVEYOR

Model 600. Quality dynamic microphone for mobile use.

Model 6002. Gain control amplifier with push-to-talk switch.

Model 6004. Noise-cancelling dynamic microphone for mobile and fixed-station use.

Scanners

E. F. JOHNSON CO.

New Pocket-Size VHF Scanning Monitor. $119.95
Automatically scans up to four VHF-FM channels. Delivers loud, clear audio over extended operating periods without battery replacement. Operates for a full 5-day week on just four alkaline penlight cells. ½ to 6 times more efficient than conventional pocket scanning monitor circuitry. Very compact. Can be used at home, office, in vehicles, or carried in a pocket.

ELECTRA COMPANY

Bearcat portable. $129.95—Light-emitting diodes, individual channel lock-out switches, 8 channels per second scanning, 250 mw RMS audio power, single function switch, belt clip, 6¼" x 1½" x 2¾", 11 oz., 6V DC.

Other models: *Jolly Roger, Bearcat III, Bearcat 101, Bearcat IV.*

JMD ELECTRONICS

Model TA. Monitor is a tone activated receiver and a monitor receiver with provisions for installing up to 11 crystal controlled channels. In alert position, receiver will monitor the alert channel and will be quiet until it receives proper alerting tones. In monitor position, any one of the 11 channels can be monitored continuously, with squelch, to hear all messages. Crystal controlled, double conversion, solid state, tone and voice receiver to give excellent reception of channel being received and suppression of unwanted channels, while using very little electric current.

Model TP pocket receiver. Tone activated pocket receiver designed to be used as a personal alerting receiver.

JMD 10 channel monitor scanning receiver. Checks each of the ten channels, one at a time, and stops at the first channel it picks up a signal on. It will stay on that channel, until it stops receiving the signal and will continue checking the channels over again, until it receives a signal on one of the channels.

MS series receivers. Are multiple channel, crystal controlled, solid state receivers. Allows you to install up to 11 crystals in the single band receiver or 17 crystals in the dual band receiver, allowing user to monitor police, fire, public safety stations, etc. by selecting the channel he wishes to listen to. Squelch control.

SURVEYOR

VHF 2 band scanner. Model 10HLP. Programmable switch. Frequency range—Lo VHF 30–50 mhz. Hi VHF 147–174 mhz. Modulation—5khz, 6 lbs. 50 ohm external antenna, audio output power—2 watts into 8 ohms.

VHF/UHF 3 band scanner. Model 10P. Frequency range, Lo VHF 30–50 mhz. Hi VHF 147–174 mhz. UHF 450–520. Audio output—2 watts into 8 ohms. 50 ohm external antenna. DC 13.5V, AC117V.

Model 10-4VHA and *Model 10-4 UHF, 4-channel pocket scanners.* Ultra compact size—5⅜" x 2½-¾" x 1½-¾" D overall. Automatic or Manual Channel Section. LED channel indicators for low battery drain. 4 crystal controlled channels with lockout switches. Built-in speaker. Adjustable volume and squelch control. Rechargeable circuitry.

TENNELEC, INC.

Model TN-800. 8 channels. Any combination of high, low and UHF. For in-home or in-car listening. Mobile bracket, DC and AC cords included with switchable scan delay. Antenna input. Matched for optimum response with 18″ whip on all bands. 24 transistors, 4 integrated circuits, 22 diodes, 8 LED indicators.

Memoryscan MS-2. $339.95. New super-selective filter screens out overlap. Exclusive visual frequency verification. New FET RF stages. 16 channels with push-button programming. Tennelec Memory Bank is "control." No crystals required, ever. Selects from more than 4,000 frequencies.

The Tennelec portable table top scanner. Model TN-450(UHF) 452–462 mhz and Model TN-150 (VHF) 151–163 mhz. Lightweight. Small. Solid state. 4 channel reception. Can be used in the home. Office. Out-of-doors. Auto. Battery operated. 6 Penlite size AA batteries used. $99.95

UNIMETRICS

Dura-Scan. $109.95. 4-channel crystal-controlled VHF/FM Automatic/Manual Scanner Monitor. Scans up to 4 crystal-controlled marine channels and "locks in" when voice signal is present. All solid-state circuit, pushbutton switches for automatic or manual scanning, modular construction for easy service if needed. Ultra-compact for hard-to-get-at spots.

Transceivers

CRAIG CORPORATION

Mobile Transceiver Model 4101. Adjustable

squelch, built-in automatic noise limiter and voice compression circuits, 23 position channel selector and LED modulation indicator. The press-to-talk switch on 4101's dynamic microphone activates an On the Air light to show you are transmitting.

Standard Mobile Transceiver Model 4102. Easy-to-read illuminated channel indicator, dual function meter and LED, Light Emitting Diode modulation indicator, to let you know when you are on the air. Includes Quick-Release anti-theft mounting, which lets you easily remove or transfer the unit.

Custom Mobile Transceiver Model 4104. Automatic noise limiter and noise blanking to cut out interference, plus illuminated channel indicator and dual function S/RF meter. Quick-Release mounting.

Base Citizens Band Transceiver Model 4201. Two-way power lets you use the 4201 as a base or mobile station. Precise crystal-controlled transmission and reception. Three-position ANL/NB/PA switch to provide automatic noise limiting and noise blanking to deal with different types of interference. Separate, built-in easy-to-read meters measure modulation, standing wave ratio, RF power output and receive signal strength.

Deluxe Mobile/Base Transceiver Model 4103. Exclusive Quick-Release anti-theft mounting system. Easy to use this model in another car, boat, or RV by the addition of optional extra Quick-Release brackets. Delta Tune for clear reception of off-frequency transmissions, RF gain control to adjust for weak signals and prevent blasting from nearby transceivers and switchable ANL plus noise blanker to eliminate interference. Built-in compression circuitry provides maximum range and voice readability. Backlighted easy-to-read 23 position channel indicator and three-way SWR/Signal Strength/Relative Power Output meter with calibration control insures maximum operational efficiency.

E. F. JOHNSON CO.

Messenger 123SJ. With solid-state meter read-out! $169.95. Channels: 23-Transmitter: Max. FCC permitted power (4 watts output). FCC type accepted-DOC approved-Size $2^1/_2$"h. x $6^3/_{16}$" w. x $9^5/_8$"d. Power required: 12 volts DC (positive or negative ground). Accessories, including portable power pack, AC base station power supply on back cover. Supplied complete with microphone, mounting bracket, and instructions.

Messenger 130A mobile. $199.95 and Messenger 132 base—$259.95. Offers private listening with automatic speaker silencing as you lift the handset. With increased clarity over background distractions of traffic or office noise. And the option of simultaneous speaker/handset operation when you want "conference call" listening. Both radios provide other deluxe features including built-in speech compression, all 23-channels and built-in PA functions.

Viking 352-SSB/AM mobile radio. Easy to use individual controls for all functions, eliminating tandem-type dual function controls. Mode selector employing color keyed lights to give indication of AM, USB, and LSB modes at a glance. Built-in PA function, illuminated metering, and dual polarity operation from positive or negative ground 12 volt DC.

Messenger 121A. $99.95. Built-in automatic speech compression on transmit and automatic noise limiting on receive. Maximum FCC permitted power. Easy push-button selection of up to five channels. Comes with microphone, mounting bracket, DC power cord and built-in loudspeaker. An accessory power supply is available to convert the radio to 110 volt AC base station operation.

GEMTRONICS

Cherokee Model GTX-2325. $369.95. SSB. RF attenuator. P-S meter, noise blanker, PA system, transmit indicator, extension speaker plug-in.

Warrior Model GTX-2300. $269.65. For base and mobile use. PA system, external monitoring, 23 channels, 5 watts.

The Apache Model GTX-23. $169.95. Three-position delta tuner, RF-S meter, squelch control, PA/CB switch, modulation lamp and noise limiter.

Comanche Model GTX-36. $139.95. Simple to operate. 5 watt power input, transmission light, PA feature, PT mike, adjustable mounting bracket.

LAFAYETTE RADIO ELECTRONICS CORP.

Dyna-Com 3b with 7-stage transmitter. $74.94. 3 crystal-controlled channels, superhet receiver, external mike/speaker jack, ceramic IF filter, combination battery/RF power and signal-strength meter.

Dyna-Com 23. $159.95. Full 5-watt 23-channel hand-held two-way radio. Frequency synthesis for 23-channel crystal control. Automatic compression range boost for greater talk power. 455 khz mechanical filter for excellent selectivity. Jack for external base station or mobile antennas. Combination battery/RF output and "S" meter. Variable squelch control plus automatic noise limiting. External microphone-speaker jack.

HB-23A 23-channel capability crystal-controlled mobile two-way radio. $139.95. 455 khz mechanical filter for razor-sharp selectivity, dual conversion receiver with RF stage, ANL, 5-watt input power with "Range Boost" circuitry, push-to-talk mike, PA speaker/earphone jack.

Telsat 1023-23 23-channel base station with all CB crystals. $179.95. Dual-conversion superhet receiver, delta-fine tune for off-frequency stations, for 117 VAC base operation, variable squelch and switchable ANL, 12-VDC input for emergency battery operation, Range-Boos circuit for greater effective range.

Telsat SSB-50A two-way radio. $329.95. 23 channels AM, 23 Upper Sideband, and 23 Lower Sideband. Suppressed sideband carrier offers great effective range and talk power, "Range Boost" circuit and automatic modulation control for 100% power, dual-conversion superhet receiver, multi-stage RF noise silencing, output for remote PA speaker, 12 VDC negative or positive ground.

PATHCOM, INC.

Model CB 2300. 23 channel synthesized design, shaped audio response for best voice clarity, built-in noise limiter for surpressing external noise, thus providing clean, clear communications, superior series gate impulse noise limiting, for processing very weak signals, full size "S" meter to monitor incoming signal strength, cascaded amplified AGC system to eliminate noise interference, front panel control for built-in PA and Loud Hailer system, external speaker/hailer jacks, local/distant control for range adjustment, 12 V DC plus or minus ground, locking mounting bracket, ceramic plug-in mike.

CB 123A. 23 channel synthesized design, S/RF meter for monitoring incoming signal strength and relative power output, transmit and receive mode indicator lights, local/distant control, PA circuitry with front panel control.

CB 133. External speaker jack, transmit mode indicator light, squelch control for quiet standby, dynamic plug-in mic.

CB 76. 23 channel synthesized design, 117V AC, lighted S meter for monitoring incoming signal strength, local/distant control to increase sensitivity or eliminate strong interference signal, cascaded amplifier ACG system provides constant audio output and lowest noise interference, external jacks for speakers or headphones, TVI shielding.

CB 1000B. 23 synthesized channels AM with 4 watts maximum legal output power, 46 channels SSB with 12 watts PEP maximum legal output power, transmit and receive mode indicator lights, AM, LSB and USB mode status indicator lights, switchable noise blanker circuit for suppressing impulse noise interference, built-in PA/CB control, full range RF gain control for receiver sensitivity adjustment and to prevent signal overload, clarifer to ensure exact tuning and voice naturalness and to adjust the clarity of the received signal, front panel plug-in mic, PA jack, S/RF meter for monitoring incoming signal strength and relative power output, SWR meter for monitoring antenna system performance, digital clock and timer/alarm switch, mounting bracket.

SURVEYOR

Model 1500. 5-watt, 3-channel transceiver with squelch control. Hand-held. Low power switch lets you transmit 2.5 watts for longer battery life, bright LED power indicator shows battery condition when set is turned on, streamline push-to-talk key flushed with side of cabinet for protection against transmitting or receiving accidentally, isolated chassis prevents circuitry damage when unit is used in 12-volt cars or trucks.

Model 2300. LCL Distributor Switch for maximum utilization of the Sensitivity when in a low-noise area. When in a high-noise area, switch to LCL position. This limits the noise and operation is more quiet. PA/CB Switch-Can be used as a public address system by pushing the PA/CB Switch to the PA position. Plug in public address speaker to speaker jack marked PUB on rear panel. External Speaker-For remote listening to receiver system when you are out of car. Backlighted 23-position channel of operation. "S" meter and modulator indicator registers incoming signal strength. Receiving indicator lights lets you know your set is receiving a signal. Transmitter indicator light will glow red when ready to transmit.

TRAM/DIAMOND CORPORATION

Model D201. VOX including front panel sensitivity and delay control, noise blanker, SWR Bridge, transmitter tone control (TTC), Mic gain control, receiver tone control, Astatic GD104 microphone supplied, fully tunable receiver, fully crystal controlled, receiver, clarifier control functional; both transmit and receiver, S/RF meter, squelch control, crystals supplied for all 23 channels.

Model D-40. Microphone gain control insures 100% modulation under all operating conditions, dependable automatic "Fail Safe" final transistor protection, SWR readout to constantly check the efficiency of your antenna, "Diamond O.C."- rigorous testing of each Diamond 40, instant positive or negative ground with polarized DC power supply, multi-function meter for power output, SWR or S functions, built-in public address capability.

Model XL5 AM/SSB. Efficient RF type noise blanker, clarifier, squelch, illuminated "S" meter, and dynamic PTT microphone. Channels: 23AM, 23LB, 23USB.

UNIMETRICS, INC.

Marlin-L-5-watt, 23 channel, solid-state transceiver. Rugged for marine use, too. Includes crystals for all 23 channels, S/PRF meter to indicate signal-strength on receive and RF output on transmit. PA system. Watertight construction and corrosion-resistant materials used throughout.

Dolphone. $214.95. 5-watt, 23-channel, all solid-state transceiver with telephone-type handset. Lets you communicate under noisiest conditions, motor noise, rough waters, etc. Also lets you receive calls in complete privacy. Squelch control, automatic noise limiting, speaker/handset switch, PA switch. For 12 VDC negative ground, adaptable for positive ground.

Stingray-I. $374.95-23 AM plus 46 Single Sideband (upper and lower) crystal controlled channels, burglar alarm switch, front panel S/PRF meter indicates signal strength on receive, RF output on transmit, RF noise cancelling circuitry, watertight, corrosion resistant, external speaker/headphone jack, built-in PA facilities, range boost and automatic modulation control, optional AC power supply.

TELEX COMMUNICATIONS, INC.

Model CB-12 Headset. $59.95

The headset combines a dynamic receiver, a ceramic boom microphone with FET amplifier and talk-switch that provides clear, intelligble sound quality for receiving and transmitting. Single muff design lets you monitor your radio at all times yet keeps you alert to the sounds around you in traffic or at the base. Adjustable headband. Padded ear cushion. You can listen without disturbing anyone else. Battery power is used only during transmitting and is readily available and simple to replace. Weighs 15 oz.

APPENDIX B
Manufacturers' Addresses

Antenna, Incorporated
23850 Commerce Park Road
Cleveland, Ohio 44122

The Astatic Corporation
Conneaut, Ohio 44030

Craig Corporation
921 West Artesia Blvd.
Compton, California 90220

CTS Knights, Inc.
Sandwich, Illinois 60548

E.F. Johnson Company
Waseca, Minnesota 56093

Electra Company
Cumberland, Indiana 46229

GC Electronics
Division of Hydrometals, Inc.
400 South Wyman Street
Rockford, Illinois 61101

Gemtronics
Division of Gem Marine Products
Box 1408
Lake City, SC 29560

Gold Line
25 Van Zant Street
E. Norwalk, Connecticut 06855

Hellstar
1600 N. Chestnut
Wahoo, Nebraska 68066

JMD Electronics
P.O. Box 346
West Hempstead, New York 11552

Lafayette Radio Electronics Corporation
111 Jericho Tpke.
Syosset, New York 11791

Leader Instruments, Corp.
151 Dupont Street
Plainview, New York 11803

Mosley Electronics, Inc.
4610 N. Lindbergh Blvd.
Bridgeton, Mo. 63044

Mura Corporation
50 So. Service Road
Jericho, New York 11753

National Radio Company, Inc.
111 Washington Street
Melrose, Mass. 02176

New-Tronics Corporation
15800 Commerce Park Drive
Brook Park, Ohio 44142

Pathcom, Inc.
Pace Communications Division
24049 South Frampton Avenue
Harbor City, California 90710

Pal Electronics Co.
Division of Fire Communications Corp.
2962 W. Weldon
Phoenix, Arizona 85017

Regency Electronics, Inc.
7707 Records Street
Indianapolis, Indiana 46226

Saxton Products, Inc.
215 N. Route 303
Congers, New York 10920

Shure Brothers, Incorporated
222 Hartrey Avenue
Evanston, Illinois 60204

Sonar Radio Corporation
73 Wortman Avenue
Brooklyn, New York 11207

Surveyor Manufacturing Corporation
29245 Stephenson Highway
Madison Heights, Michigan 48071

Telex Communications, Inc.
9600 Aldrich Avenue South
Minneapolis, Minnesota 55420

Tennelec, Inc.
Commercial Products Division
Oak Ridge, Tennessee 37830

Tram Corporation
Lower Bay Road
Winnisquam, N.H. 03289

Unimetrics, Inc.
1534 Old Country Road
Plainview, New York 11803